DevOps

DevOps

by Emily Freeman

FOREWORD BY Nicole Forsgren

DevOps For Dummies®

Published by: **John Wiley & Sons, Inc.**, 111 River Street, Hoboken, NJ 07030-5774, www.wiley.com

Published simultaneously in Canada

For general information on our other products and services, please contact our Customer Care Department within the U.S. at 877-762-2974, outside the U.S. at 317-572-3993, or fax 317-572-4002. For technical support, please visit https://hub.wiley.com/community/support/dummies.

Wiley publishes in a variety of print and electronic formats and by print-on-demand. Some material included with standard print versions of this book may not be included in e-books or in print-on-demand. If this book refers to media such as a CD or DVD that is not included in the version you purchased, you may download this material at http://booksupport.wiley.com. For more information about Wiley products, visit www.wiley.com.

Library of Congress Control Number: 2019945514

ISBN: 978-1-119-55222-2

ISBN 978-1-119-55223-9 (ebk); ISBN 978-1-119-55224-6 (ebk)

10 9 8 7 6 5 4 3 2 1

Contents at a Glance

Table of Contents

Foreword

W*hat is DevOps?*

This question is one of the most common questions I get in my work — and I hear it from experts and novices alike. I've worked in technology for almost two decades and have been a DevOps researcher, strategist, and expert who has guided hundreds of technology leaders and engineers to make their software better, allowing them to deliver value to their customers faster and safer. Yet, many of us in this field still hear this question.

Unfortunately, a single, universally accepted definition of DevOps doesn't exist — much to everyone's chagrin. But if we're honest, it probably doesn't matter; after all, having one clear definition codified in the Agile Manifesto didn't help the Agile community much, either.

With *DevOps For Dummies,* Emily Freeman has written a book about DevOps that you can read from cover to cover as a book, or use as a reference, or jump around in for a choose-your-own adventure exploration through DevOps concepts. It's a brilliant way to structure the content because DevOps covers some development, some operations, and a whole lot of culture — plus a whole lot more. In this book, Emily has done a great job of digging into important concepts for teams implementing this new way of work, whether you're greenfield or brownfield or have no idea of what the words *greenfield* or *brownfield* mean.

Emily brings clear eyes and a fresh voice to the topic, crafting insightful narratives and breaking down concepts into clear writing. By coming into technology mid-career, she brings an understanding and comprehension that others who were "born" into the tech sometimes take for granted. Her writing communicates these (sometimes hidden) details effortlessly, walking the reader through the landscape with ease and wit.

Some of my favorite sections are those on developing code so that it's clear and maintainable (check out Chapter 9, especially the part about peer review) and empowering teams to help you scale. I have too many favorite chapters and sections to list, so I urge you to find your own favorites and not to run out of sticky notes and highlighters!

I wish you the best of luck on your DevOps journey. Remember that where you start doesn't matter; what matters is that you keep going and keep improving.

— Nicole Forsgren, Research & Strategy at Google Cloud and co-founder and CEO of DevOps Research and Assessment (DORA)

Introduction

I believe that the greatest challenges facing the tech industry aren't technical; they're human. Think about it: Hardware and computing are more powerful than they ever have been. Automated tools remove the drudgery of rote work. Robust frameworks and libraries build shortcuts and functionality into applications for you. You can do more, faster, than ever before.

The problem that organizations face now is that of the social dynamics of humans on engineering teams. Specifically, those dynamics are the natural friction that arises from misaligned incentives and poorly communicated goals; the frustration in attempting to explain a concept or approach to someone who has a different expertise than you; and the fear people feel when they think about looking stupid in front of their colleagues or being automated out of a job.

DevOps addresses all these issues, and this book explains how.

About This Book

I've designed this book to be a resource that someone who has never heard of DevOps — or doesn't know what it actually means — can walk through to gain a broad understanding of DevOps and how it fits into the software development life cycle (SDLC) — that is, the entire process of how software is developed and released. Although SDLC has traditionally used the word *development,* I prefer *delivery* because it removes the possible elevation of developers over other disciplines.

I've ordered the information in this book to be both approachable chronologically as well as selectively. You can read it straight through and allow each section to build on the last or you can hop around to your heart's content.

I use the following conventions throughout the book:

» Web addresses and programming code appear in `monofont`. If you're reading a digital version of this book on a device connected to the Internet, you can click or tap the web address to visit that website, like this: `www.dummies.com`.

» New terms that I define appear in *italics*.

» This book uses *they* (and its derivatives) as a gender-neutral, singular pronoun throughout.

Foolish Assumptions

When I first started writing this book, I struggled to identify my main audience. Developers? Operations people? Executives? It was important to me to make DevOps approachable and real. Too often, people talk about DevOps only in the context of greenfield projects and companies with nearly endless resources. I wanted to pull back the shine and get to the substance of DevOps as a discipline — one that helps engineers do their job better and faster. I also wanted to ensure that I met the reader where they were and spoke to people who had never heard of DevOps as much as I spoke to the engineers who are well on their way to advanced DevOps practices. In the end, I focused on anyone who needs to deliver business priorities as much as retain engineering talent. They need realistic solutions to real-world problems. They see the benefits of automation but also need to ensure security and compliance with regulatory bodies.

Regardless of whether you fit that exact profile, I hope that you can glean what you need from this book and that it may play a small part in your success as you evolve and improve your engineering practice.

Icons Used in This Book

The Tip icon marks tips (duh!) and shortcuts that you can use to make implementing DevOps practices easier.

Remember icons mark the information that's especially important to know. To siphon off the most important information in each chapter, just skim through these icons.

The Technical Stuff icon marks information of a highly technical nature that you can normally skip over.

The Warning icon tells you to watch out! It marks important information that may save you headaches and unnecessary conflict.

Beyond the Book

This section describes where readers can find book content that exists outside the book itself. A *For Dummies* technical book may include the following, although only rarely does a book include all these items:

- **Cheat Sheet:** You can find the Cheat Sheet for this book by searching its title at `www.dummies.com`.
- **Updates:** If this book has any updates, you can find them on this book's page at `www.dummies.com`.

Where to Go from Here

I've written this book in such a way that you can skip around if you so choose. If you're a developer, you may find that you know most of the information in Part 2, which covers the development pipeline (though I suggest skimming it to catch a few new ideas!). If you're an operations engineer, you may feel more confident in some of the more infrastructure-focused chapters.

A DevOps transformation is no small feat, nor is it an overnight process. It will take hours of planning, honest conversations, brainstorming, reflection, and technical changes. Don't rush through the process. Simply learning and thinking about your everyday work from a different perspective is a healthy way of waking up your mind. The journey is just as valuable as the outcome.

1 Demystifying DevOps

IN THIS PART . . .

Understand DevOps values and priorities, which focus on people, process, and technology.

Design your organization's culture with DevOps in mind through encouraging teamwork, reducing silos, and embracing failure.

Identify waste and locate bottlenecks along your software development life cycle to locate the easiest (and most immediate) wins for your DevOps transformation.

Persuade your colleagues, from executives to engineers, of the benefits of DevOps for an engineering organization.

Measure your work and track your DevOps successes, allowing everyone to see the incremental improvements.

» Understanding DevOps values

» Seeing how your organization benefits

Chapter 1

Introducing DevOps

DevOps has transformed the way engineering teams collaborate to create and ship software. It's a broad and encompassing philosophy that inspires diverse implementations across the industry.

I define DevOps as an engineering culture of collaboration, ownership, and learning with the purpose of accelerating the software development life cycle from ideation to production. DevOps can enable you to reduce interpersonal friction, eliminate bottlenecks, improve collaboration, increase job satisfaction through engineer empowerment, and accelerate team productivity. DevOps is no silver bullet, but it can have massive impact on your organization and your products.

In this chapter, I emphasize the importance of culture over process and tooling, discuss the principles and values of DevOps, and dive into how your organization will benefit from a DevOps approach.

What Is DevOps?

This book has no exact DevOps prescription for you — because none exists. DevOps is a philosophy, one that that prioritizes people over process and process over tooling. DevOps builds a culture of trust, collaboration, and continuous improvement. As a culture, it views the development process in a holistic way, taking into account everyone involved: developers, testers, operations folks, security, and infrastructure engineers. DevOps doesn't put any one of these groups above the others, nor does it rank the importance of their work. Instead, a DevOps company treats the entire team of engineers as critical to ensuring that the customer has the best experience possible. (You can find more about company culture in Chapter 2.)

DevOps evolved from Agile

In 2001, 17 software engineers met and published the "Manifesto for Agile Software Development," which spelled out the 12 principles of Agile project management (see the sidebar "The origins of Agile" in Chapter 7 for more details). This new workflow was a response to the frustration and inflexibility of teams working in a waterfall (linear) process. Working within Agile principles, engineers aren't required to adhere to original requirements or follow a linear development workflow in which each team hands off work to the next. Instead, they're capable of adapting to the ever-changing needs of the business or the market, and sometimes even the changing technology and tools.

Although Agile revolutionized software development in many ways, it failed to address the conflict between developers and operations specialists. Silos still developed around technical skill sets and specialties, and developers still handed off code to operations folks to deploy and support.

In 2008, Andrew Clay Shafer talked to Patrick Debois about his frustrations with the constant conflict between developers and operations folks. Together, they launched the first DevOpsDays event in Belgium to create a better — and more agile — way of approaching software development. This evolution of Agile took hold, and DevOps has since enabled companies around the globe to produce better software faster (and usually cheaper). DevOps is not a fad. It's a widely accepted engineering philosophy.

DevOps focuses on people

Anyone who says that DevOps is all about tooling wants to sell you something. Above all else, DevOps is a philosophy that focuses on engineers and how they can better work together to produce great software. You could spend millions on every DevOps tool in the world and still be no closer to DevOps nirvana. Instead, focus on your most important engineering asset: engineers. Happy engineers make great software. How do you make happy engineers? Well, you create a collaborative work environment in which mutual respect, shared knowledge, and acknowledgement of hard work can thrive. See Chapters 2 and 15 for more about how to create teams of happy, empowered engineers who embody a growth mindset and take pride in their work.

Company culture is the foundation of DevOps

Your company has a culture, even if it has been left to develop through inertia. That culture has more influence on your job satisfaction, productivity, and team velocity than you probably realize.

Company culture is best described as the unspoken expectations, behavior, and values of an organization. Culture is what tells your employees whether company leadership is open to new ideas. It's what informs an employee's decision as to whether to come forward with a problem or to sweep it under the rug.

Culture is something to be designed and refined, not something to leave to chance. Though the actual definition varies from company to company and person to person, DevOps is a cultural approach to engineering at its core.

A toxic company culture will kill your DevOps journey before it even starts. Even if your engineering team adopts a DevOps mindset, the attitudes and challenges of the larger company will bleed into your environment.

With DevOps, you avoid blame, grow trust, and focus on the customer. You give your engineers autonomy and empower them to do what they do best: engineer solutions. As you begin to implement DevOps, you give your engineers the time and space to adjust to it, allowing them the opportunities to get to know each other better and build rapport with engineers with different specialties. Also, you measure progress and reward achievements. Never blame individuals for failures. Instead, the team should continuously improve together, and achievements should be celebrated and rewarded.

You learn by observing your process and collecting data

Observing your workflow without expectation is a powerful technique to use to see the successes and challenges of your workflow realistically. This observation is the only way to find the correct solution to the areas and issues that create bottlenecks in your processes. Just as with software, slapping some Kubernetes (or other new tool) on a problem doesn't necessarily fix it. You have to know where the problems are before you go about fixing them. As you continue, you collect data — not to measure success or failure but to track the team's performance. You determine what works, what doesn't work, and what to try next time. In Chapter 3, you learn how to identify bottlenecks in your development process.

Persuasion is key to DevOps adoption

Selling the idea of DevOps to your leaders, peers, and employees isn't easy. The process isn't always intuitive to engineers, either. Shouldn't a great idea simply sell itself? If only it were that easy. However, a key concept to always keep in mind as you implement DevOps is that it emphasizes people. The so-called "soft skills" of communication and collaboration are central to your DevOps transformation. Persuading other folks on your team and within your company to adopt DevOps requires practicing good communication skills. Early conversations that you have with colleagues about DevOps can set you up for success down the road — especially when you hit an unexpected speed bump.

Small, incremental changes are priceless

The aspect of DevOps that emphasizes making changes in small, incremental ways has its roots in lean manufacturing, which embraces accelerated feedback, continuous improvement, and swifter time to market. When I talk about DevOps transformations, I like to use water as a metaphor. Water is one of the world's most powerful elements. Unless people are watching the flood waters rise in front of them, they think of it as relatively harmless. The Colorado River carved the Grand Canyon. Slowly, over millions of years, water cut through stone to expose nearly two billion years of soil and rock.

You can be like water. Be the slow, relentless change in your organization. Here's that famous quote from a Bruce Lee interview to inspire you (`https://www.youtube.com/watch?v=cJMwBwFj5nQ`):

Be formless, shapeless, like water. Now you put water into a cup, it becomes the cup. You put water into a bottle, it becomes the bottle. You put it in a teapot, it becomes the teapot. Now, water can flow or it can crash. Be water, my friend.

Making incremental changes means, for example, that you find a problem and you fix that problem. Then you fix the next one. You don't take on too much too fast and you don't pick every battle to fight. You understand that some fights aren't worth the energy or social capital that they can cost you.

Benefitting from DevOps

This entire book dives into how you and your team can benefit from implementing DevOps in your organization. Beyond the human component, which enables faster delivery, improved functionality, and fearless innovation, DevOps has technical benefits.

Continuous integration and continuous delivery (CI/CD) are closely aligned with DevOps. Continuous software delivery removes many of the bottlenecks often seen in teams that deploy infrequently. If you create automated pipelines that pass new code through a robust test suite, you can feel more confident in your deployments. (I talk more about CI/CD in Chapter 11.)

DevOps also enables faster recovery from incidents. You will inevitably experience a customer-impacting service disruption at some point, no matter how well tested your code is. But teams who work in a DevOps methodology find resolutions faster through better coordination, more open accessibility, shared learning, and better performance monitoring.

Engineering is not the only side of your organization that benefits from DevOps. The business side of your organization will see fewer customer complaints, faster delivery of new features, and improved reliability of existing services.

DevOps enables you to do more with the resources you already have. It accepts the reality of constraints and shows you how to succeed within your unique environment.

Keeping CALMS

As you begin to get familiar with DevOps, you'll likely come across a model called CALMS. It stands for culture, automation, lean, measurement, and sharing, and it's a helpful framework through which to understand the DevOps principles and evaluate your DevOps success as you apply those principles throughout your organization.

Culture

Your culture needs to be collaborative and customer centered, which means your engineers understand that the purpose of technology is to make your customers' lives easier. If customers don't find value in the product, the product will fail. Technology is secondary to this goal. The best DevOps cultures are extremely collaborative and cross-functional, with people from different teams and varying skill sets working together to engineer a better product. Listening is a major component of communication, and an easy litmus test of culture is to listen to conversations. Are people constantly talking over each other? If so, I bet you have opportunities for major cultural improvements ahead.

Automation

Rote tasks are an engineer's worst nightmare, not only because they're, well, boring, but because they're inefficient. Engineers speak computer so that they can make computers do the jobs that people don't want to do. Usually the lowest-hanging fruit for improvements in automation are code builds, automated testing, deployments, and infrastructure provisioning. I dig deeper into identifying low-hanging fruit in Chapter 3.

Lean

Lean doesn't refer just to lean manufacturing. It applies more widely to the nature of DevOps teams, which are agile and scrappy. Lean teams eschew low-impact activity because it doesn't provide value to the customer. Another aspect of lean is how it keeps to the goal of continuous improvement. Everyone embraces a growth mindset and earnestly wants to improve.

Measurement

Data is critical to DevOps. Measuring progress through data will inform nearly every aspect of your organization's transformation. Keep in mind, though, that progress should never be tied to individual performance. Think of it as tracking your progress along an endless marathon rather than as a way of knowing when you're "done." You're never done. No one is.

Instead of regarding the data you collect as a measure of how poorly you're doing, think of it as gauging your improvement. Celebrate the wins. That approach bolsters the entire team and keeps your engineers happy, motivated, and productive. I guarantee you're doing some things well, and highlighting the good is important. I talk about what you can measure in Chapter 5.

Sharing

DevOps was founded because operations and development had some conflict. They lacked common ground and were incentivized based on different standards. Operations folks are typically measured on the reliability and availability of an application, whereas developers are, more often than not, incentivized to create new features for the application. (I talk more about how operations and development are measured in the next section.) You know what the biggest threat to uptime is? Deployments. Developers initiate deployments with new code releases. Thus, operations folks hate developers. Now, it's not usually that bleak, but there's a seed of truth in that. The friction makes solving problems nearly impossible and turns everything into a blame game. DevOps seeks to change that atmosphere completely and create an environment in which both teams teach each other and feel empowered — thus building a single team through which everyone contributes.

AN ENGINEER'S TALE: WHAT DROVE ME TO DEVOPS

I want to let you in on a little secret. I came to DevOps by accident. Yep! Totally an accident. But I think my story speaks volumes about the power of the DevOps movement and community.

I was a backend Java engineer at a small company with a traditional engineering team. The team consisted of a dozen developers and two operations folks. (Sounds to be about the usual ratio, right?)

The code had a bug. I updated the code that selected preview images in the application. Yet, the home page wasn't displaying the changes, and ops blamed me. I looked into it and concluded that it was a content delivery network (CDN) issue. Because of access constraints of the developers on my team, I couldn't mitigate the problem myself. I needed the ops team.

The ops expert felt that this was a code issue and refused to help me. We went back and forth three times before I went into a closet and angrily typed an abstract. *Humpty Dumpty: A Story of DevOps Gone Wrong* was my first tech talk and was inspired by my personal experiences and frustrations with developers facing off against operations folks.

In that company, and so many others, the operations team was a bottleneck. They prevented me from doing my work. It wasn't their fault, though. The people involved highlighted the problem, but the problem itself was a process issue.

My experience at that job led me to DevOps, which piqued my interest. In the course of learning about DevOps, I found incredible relief in the discovery that the issues I had encountered weren't about just me! I wasn't a bad developer. I was just a human, and other engineers felt the same frustrations in their jobs. It is my greatest wish that this book can both reassure you that your experience is valid and common as well as show you some approaches that can make your job just a little bit more awesome.

Solving the problem of conflicting interests

On traditional engineering teams, developers (those who write the code) and operations engineers (those who deploy systems and maintain infrastructure) are on opposing sides of a never-ending war. Okay, that's not exactly accurate. But they don't get along, and that's because they're measured by different criteria.

Developers are typically measured by the number of features they release or the number of bugs they fix. (Evaluating developers by lines of code written is a terrible idea. Many times, the best developers delete more lines of code than they add.)

Unfortunately, code quality and reliability aren't typically measured. As a result, developers naturally prioritize the work that will make them look productive. They don't spend time refactoring code to make it more readable or paying off technical debt accrued from the last big product push.

In contrast to how developers are measured, operations teams are typically measured by site reliability and uptime. You've likely heard of the five 9s: 99.999 percent availability. The five 9s means that your site can be down for only five minutes per year. Five minutes . . . *per year.* That's a lot to ask. It's also expensive to maintain because of the number of storage and compute resources you must have at your disposal, not to mention the personal impact it has on the operations individuals tasked to keep availability at that level. Those people are often asked to take on heroic efforts and respond to problems regardless of the day, time, existing workloads, or personal obligations.

To make the conflict clear: In traditional engineering organizations, developers must deploy new code to release new features. But deploys are the most common action that initiates service disruptions and site outages.

Two problems come from this situation:

- **Responsibility is siloed.** Developers don't know how to release or support their code, and they lack systems knowledge that enables them to understand infrastructure requirements. Most developers don't know (or care) how their code actually runs. Their job is done.
- **The goals and incentives are in opposition.** Developers toss code over the operations team and expect them to deploy the code and ensure that it runs perfectly. Operations folks are incentivized by uptime, availability, and reliability. They often assume that the code is poorly written and they'll be yelled at (or fired) for an incident that isn't their fault.

Do you see why you hear audible sighs when developers and operations teams interact? DevOps seeks to eliminate both the challenges created by siloed responsibility and opposing goals. By aligning incentives, sharing knowledge, removing barriers, and respecting different roles, DevOps can dramatically improve the interpersonal communication and cooperation on your team.

- Understanding DevOps values
- Crafting a vision statement
- Incentivizing your values

Chapter 2 Designing Your Organization

Company culture is best described as the unspoken expectations, behavior, and values of an organization. Culture is what tells your employees whether company leadership is open to new ideas. It informs an employee's decision on whether to come forward with a problem or sweep it under the rug.

Your employees and colleagues make a thousand decisions a day — all without the help of management. (This is great! Who wants a micromanager?) Culture is what informs those small but incessant decisions, so it behooves you to make sure your company culture is one that benefits the employees and ensures that their working environment is a happy place to be.

I've worked for companies with great culture. I've also worked at places in which the tension of the environment was palpable. The difference is stunning. In the former, I performed at a higher level, thought outside the box and took risks, and was happy to stay with the company for years. In the latter, I was miserable. I started to plan lunch with colleagues at 10 o'clock and couldn't wait to take every single minute until I returned. Then I suffered through the afternoon before I could leave. I wasn't motivated. I didn't do great work. I did just enough to not get fired.

Look around you. What kind of culture do you think your company has? This chapter gives you specific ways to accurately evaluate your company's culture. Also in this chapter, you find out how to develop a vision for your company culture, apply DevOps values to your engineering teams, and incentivize and reward the values you prioritize.

Assessing Your Culture's Health

One of the greatest challenges for companies — specifically older and larger organizations — is identifying the true state of their culture. Even young companies can easily overestimate the quality of their culture. If you think you have a healthy culture, that's a great start. But look at it with a cynic's eye. It's easy to see culture through rose-colored glasses.

A few years ago, Gallup released the 2017 State of the American Workplace Report. The poll found that only 33 percent of employees felt engaged at work, and a mere 22 percent of workers believed that leadership had a clear direction for the company. Those statistics aren't exactly encouraging. Here are some ways you can start to home in on the true state of your company's culture:

- **Survey your employees.** A survey is perhaps the easiest way to evaluate the state of your company culture. You must ensure that the survey is anonymous so that employees feel free to be honest with you without fear of retaliation. Include only important questions that will reveal how your employees and colleagues actually feel.

  TIP

  The best surveys ask questions about an employee's satisfaction and happiness. These are questions like, "On a scale of 1–10, how likely would you be to leave for a 10 percent raise from another company?" and "On a scale of 1–10, how would you rate your direct supervisor's job performance?

- **Observe interpersonal communication.** You can learn a lot from simply observing how a team communicates with itself. Are colleagues spoken to with respect? Do people assume positive intent? Does everyone seem engaged in meetings? Pay close attention to disagreements. If employees are quick to generalize, name call, or escalate the conflict to anger, these behaviors can hint at an inability for people to express their frustrations in a more professional way.
- **Take a hard look at leadership.** Company culture flows down from the top. The standards and priorities set by leadership have an enormous impact on the overall culture of the company. If your CEO behaves like a jerk, chances are you have a culture of fear on your hands.

After you gain a clear view of what your company culture says in the present, you can take action and ensure that the message being sent to employees is the one you want. And don't be afraid to find out that your culture is in rough shape. Opening your eyes to the honest state of your work environment is empowering. Don't think of yourself as being at the bottom of a mountain. Instead, imagine yourself kicking off from the bottom of the ocean.

REMEMBER

Company cultures often fall into four categories: apathetic, caring, exacting and integrative:

- **Apathetic:** Very little concern is shown for people or performance.
- **Caring:** People are top priority and cared for deeply while performance issues can fall by the wayside.
- **Exacting:** The reverse of caring, this culture prioritizes performance over everything else.
- **Integrative:** High concern is shown for both people and performance. This culture is ideal because both the employees and the product can thrive.

SURVEY THE RIGHT WAY

Many years ago, a tech company sent out a company-wide survey and stated that responses would remain anonymous. Employees were free to rate the company's success in a number of areas as well as express any concerns they had. Many women in particular were brutally honest and wrote of experiencing sexual harassment — a common problem in all workplaces but especially prevalent in the male-dominated world of tech.

The company misled its employees. The survey wasn't anonymous. Instead, the results were sent directly to leadership. One C-level executive took it upon himself to interview the men named as sexual harassers and to inform the women who expressed concern that he had investigated the issue and found no wrongdoing.

Let this story serve as a warning. This incident was an egregious violation of trust, one that you should never inflict on your company. If a survey is anonymous, make it truly anonymous, because after trust is lost, regaining it is almost impossible.

Integrating DevOps

In the novel *The Phoenix Project*, Gene Kim notes the following:

> *A great team doesn't mean that they had the smartest people. What made those teams great is that everyone trusted one another. It can be a powerful thing when that magic dynamic exists.*

DevOps, above all else, is a cultural shift that empowers engineers to learn freely, share responsibility, and succeed — as well as occasionally fail — *together*. If you take only one thing away from *DevOps For Dummies*, I want it to be the list of the core values that are central to the DevOps movement, as described in the next section.

Integrating these values into your everyday workflow and overall company culture results in phenomenal impacts to engineer happiness and productivity. People begin to trust each other, and through trust, collaboration can become the norm. Only then can innovation take place.

No matter how you integrate DevOps into your company's culture, of critical importance is for you to recognize that culture is central to any DevOps transformation. DevOps is a cultural revolution that unites the traditionally adversarial sides of development and operations. It encourages teamwork, collaboration, communication, and — above all else — trust in the people with whom you work.

Establishing DevOps Values

DevOps is centered around a few core principles. In this section, I highlight what I think are the seven most important values of DevOps. Some resources you find will list fewer; others, more. Here are the values I describe throughout this section:

- Encourage teamwork.
- Reduce silos.
- Practice systems thinking.
- Embrace failure.
- Communicate, communicate, communicate.

» Accept feedback.

» Automate processes (when appropriate).

The descriptions in the following sections serve as an overview of these values. If you have questions about each one, fear not! You dive into these more deeply throughout the rest of the book. Think of this as an introduction to the heart of DevOps.

Encourage teamwork

Empower your team members to make independent decisions based on their expertise. Ideally, teams will share responsibility so that everyone is accountable with regard to both celebrations and failures. Collaboration is a core principle of any DevOps culture. It's also foundational to the practice. Without this one value, your team will struggle to adopt DevOps.

Teams must trust each other. Create opportunities for your employees and colleagues to get to know each other and build rapport. For example, if you know the birthday of your coworker's daughter, you probably have a healthy relationship, which makes struggling through product decisions and working through conflict are a lot easier. Trust is the foundation of all relationships, including in engineering.

Reduce silos

Share information freely among colleagues, teams, and skill sets. Ideally, you should build cross-functional teams in which members have varying and complementary skill sets that, together, support a single product line or software service.

You may have heard of the "Wall of Confusion," which traditionally existed between developers and operations folks. Managers used to group highly specialized developers, who engineered new features and then tossed that code over to operations to deploy and support. That approach created silos of knowledge that limited collaboration. Instead of following that tactic, you want to ensure that information is shared freely among people and departments. *Everyone* is responsible for creating and delivering great software. "It's not my job" is a phrase that should never, ever be uttered by anyone on your team.

TECHNICAL STUFF

Think of technical skill sets as being T-shaped. You're looking for developers who have deep knowledge of their area of engineering. Perhaps they're a Python engineer or a front-end engineer skilled in React. That same developer should have shallow knowledge of areas like automated testing, database storage, deployment pipelines, and infrastructure. Your ops folks are never going to be your best coders, and the reverse is also true. That's not the point of DevOps. Instead, the point is remove barriers and allow information and knowledge to be shared freely.

Practice systems thinking

View everything your engineering team touches as part of a greater whole. This holistic view gives you a better understanding of how the team functions and where you can improve. Instead of viewing the whole as a grouping of individual elements, think of the team as an ecosystem.

The human body has a circulatory system, a digestive system, and many other separate functions, but these systems and functions all work together, and all parts are necessary for survival. Your engineering team is the same. Yes, members of the team have different areas of focus and specialization, but they aren't simply the sum of their parts. The team works together like a living, breathing organism.

Embrace failure

Failure is unavoidable. It happens. And yet, you likely spend much of your time attempting to avoid failure at all costs. But failure isn't always a bad thing. In fact, small failures hint at a culture that encourages risk — trying new things and innovating. Innovating and moving quickly is impossible without a few hiccups along the way.

By embracing failure, you turn the societal pressure to avoid failure on its head. Empowered by this growth mindset, you can budget for error and integrate recovery into your feedback loop. I talk more about this loop in the Chapter 13.

The key here is to view failure as a natural part of life, as well as of the development life cycle. That way, when you're faced with an unexpected and potentially large failure, you can recover quickly and continue to innovate.

Communicate, communicate, communicate

As mentioned earlier, teamwork is crucial to DevOps, and teamwork goes hand in hand with communication. Yet communication is something that engineers tend to undervalue. Despite the general belief that communication is a "soft skill," the best engineers are those who can convey technical concepts to others clearly.

Some folks may seem to be naturally good at communication, whereas others aren't. It can seem as though some are born natural communicators and the rest are destined to struggle. But the truth is that communication is a practiced skill.

Most teams struggle to communicate well. Often, engineers miss each other, or the message isn't received as intended. Because of the impact of these communication struggles can have real impact on speed, quality, and profitability, these so-called "soft skills" are important for teams to consider and prioritize. I loathe the term because I consider the "soft" skills of communication, relationship-building, project management, and conflict resolution to be some of the hardest challenges you can take on. Still, the term encompasses a need for technical folks to better engage with one another, build rapport, and establish trust.

TIP

Communication doesn't have to take place in a meeting. Overtaxing engineers with endless meetings quickly erodes any progress you've made on your DevOps journey to that point. Instead, meet your engineers where they are. Where do they prefer to meet? What methods of communication do they prefer to use? Utilize communication tools and techniques to adapt to the team's preferred style.

Accept feedback

Feedback is a gift. It doesn't always feel like it. (Believe me, I've felt some negative feelings when receiving feedback from my editors on this book.) But feedback is what enables you to realistically study and improve your software.

You don't build software to show off your coding skills. In fact, the vast majority of users will never read the code on which you spent hundreds of hours working — even on an open source project. Your users care only about whether your product actually works. Can they check their email? Can they view their invoices? Can they pay their clients? Can they buy shoes from you? The businesses vary, but the expectations of your customers don't.

Listening to your customers is the best way to quickly identify what areas of your application need improvement. If you pay attention, you can learn a lot from customers. They will tell you what they like, what they hate, and what they want from you. If you follow up and fulfill those expectations, you'll earn their loyalty.

Automate processes (when appropriate)

Have you noticed that the most technical principle is last in this list of values? As you continue in this book, you'll notice that I deprioritize technology. Why? Because technology is the least complicated and least critical aspect of creating a DevOps culture. Improved technical practices are the result of a DevOps transformation, not the journey itself.

That said, automation is married to DevOps. (This situation is true in part because vendors have products to sell, and selling ideas is hard. But that's a different book.) Automation is a tool used to practice the values of DevOps. With automation, you develop better software faster and maintain applications with better reliability. You build, deploy, and monitor your software with automation tools to improve accuracy and eliminate manual bottlenecks.

The important part of automation is that it's employed only when appropriate, and only after you've understood and manually solved the problem. Automating a failure-ridden process only helps you fail more spectacularly and abstract the source of the failure — which makes resolution more difficult. Automation is the last step in a long process, but it is still vital to enable you to use DevOps with increasingly complex software systems.

Modeling Company Culture

Organizational structure plays an enormous role in your company's culture. At an earlier time, all companies were mostly the same because most of them were in some type of manufacturing. The manufacturing industry demanded a certain type of setup, which usually involved having some sort of boss overseeing a small group of middle managers and the (typically) men on the manufacturing floor.

Then a service economy emerged and new organizational structures began to surface, with new kinds of problems. Unfortunately, this book can't give you a silver bullet for all organizational challenges. Instead, I show you a variety of solutions to the problems you face and help you choose which might be the best solutions for you and your organization. Don't stress if you try one and it doesn't quite work out. Humans are complicated, and finding a culture that allows everyone to thrive can take some time. Your company culture will evolve and you will deal with some trial and error along the way. The following list presents four types of structures into which most companies fall. As you read, consider which one your company most resembles, and which one you prefer to work in.

- **Clan:** Think of this company culture as a family-like structure of people. This culture is most often found in early-stage startups. Colleagues are collaborative, and managers (if they exist) are dedicated to their employees. Engagement is high, but sometimes a desire for agreement and harmony may drown out dissenting opinions, making way for a homogeneous perspective to emerge.

- **Meritocracy:** In this culture, great ideas are prioritized — whether the idea comes from the CEO or the lowest-level junior engineer. This principle sounds amazing on paper, but the meritocracy isn't all sunshine and rainbows. Meritocracies don't acknowledge the natural human instinct toward hierarchy and authority bias (which means that an executive's idea is bound to be overvalued). Because of power structures, both conscious and unconscious, not all ideas are perceived as equal.
- **Holacracy:** This type of culture is as simple as company culture gets. Employees manage their work independently, with full autonomy, and the company structure is completely flat. You have no bureaucracy and no micromanagement — because you have no managers. This style of organization has mixed reviews. Some companies claim to thrive in it. Others tend to use it in the early days and then integrate more hierarchy and management into the company as it grows.
- **Traditional hierarchy:** Many people argue that the hierarchical culture is outdated. Yet, most of our organizations reflect this structure (sometimes with bits and pieces of the other structures thrown in). Often in a hierarchy, communication flows down from managers to engineers. If employees haven't been empowered by the managers, this downward flow can quickly cause employees to stop innovating and suggesting new ideas because the friction encountered is simply too high.

A new type of structure is emerging as some companies merge a flat holacracy with a traditional hierarcy. In this *flatarchy* — typically seen in startups — some management layers are eliminated to provide a flatter structure, and employees are expected to communicate ideas up the chain of command and challenge downward information flow.

What style is your company now? Do you think it's the best organizational structure in which to begin your DevOps journey? Think about what advantages you might have based on your organizational values and how employees relate to each other. Also consider your disadvantages. For example, a company with a strong management layer will likely need buy-in from managers because the engineers likely defer to their judgment. A holacracy or flatter structure, on the other hand, requires a groundswell of excitement from the engineers closest to the keyboard.

Avoiding the worst of tech culture

Tech culture hasn't received the best press in recent years. Multiple scandals at numerous large tech companies have not put those companies in a particularly flattering light.

The culture of engineering is casual and centered around intelligence. So casual, in fact, that in the greater population, developers are better known for their hoodies and jeans than for their great code.

Traditional tech culture is known to be composed of engineers who are male, pale, and overworked. (Perhaps a little curmudgeon-y, too!) The tech landscape is changing, however, and DevOps is leading the way toward a more balanced and diverse engineering culture. The following tips can help you avoid some of the worst tech scandals in recent years and instead build a company known for its happy and productive employees — not to mention great software:

- **Demand diversity.** Social diversity — differences of age, race, religion, sex, and sexual orientation — is vital to producing great products and ensuring a welcoming environment. Engineers who are passionate about DevOps appreciate and encourage social diversity as well as diversity of experience and skill sets because all these features increase innovative thinking and successful problem solving. Social diversity helps to guarantee that your software is free of unconscious bias. This diversity is even more critical for companies working in machine learning (ML), artificial intelligence (AI), and big data.
- **Ensure that employees go home at a reasonable hour.** It's such a simple gesture. Make sure that your employees don't work more than 40 hours a week. Yes, your engineers know that if the site goes down, they might not make it home for dinner. Those situations should be few and far between, however. "Off hours" work and deployments are unnecessary and a symptom of larger challenges within your organization. Engineering work is incredibly taxing, and breaks are absolutely required to avoid burnout. That means restful weekends, evenings free of texts and emails, and laptop-free vacations.
- **Provide great insurance and other benefits.** If you're in the United States, you know how much medical insurance matters to your employees and their families. Many other benefits also help keep your engineers healthy and happy. Engineers are disproportionately affected by anxiety and depression. Provide opportunities for employees to improve their mental health, such as through therapy, yoga, exercise, or anything else. Give them the time (and if you can manage it, the money) to pursue activities to keep themselves healthy — physically and mentally.
- **Encourage alternative thought.** Creating a diverse and inclusive environment to work in requires that all ideas and perspectives are welcome. This diversity goes beyond what people look like and instead pulls from their experiences, stories, and perspectives. DevOps emphasizes creative problem solving, which means that you have to create space for people to share ideas — even if they're a little out of the box. In this way, junior engineers are sometimes even more valuable than your senior engineers because they bring a raw and drastically different perspective.

Crafting your vision

You may be wondering how a vision differs from a mission statement. I like to think of a vision as being the keynote of your mission. It sets out the ambitious goals of your organization. A vision is inspirational and meant to unify people behind a single, focused idea. The mission statement then fills in the gaps with a more detailed strategy and idea.

Your vision is how you pull together a focused goal for your company culture. It is the most ambitious view of where you would like to see your company go — for customers, employees, and stakeholders. It should reflect the principles of the founders and evolve as the company grows.

Ultimately, culture is what allows employee enthusiasm to thrive. Your vision statement will focus you and inform decisions for you and everyone who works for the company. You can consider it a beacon, calling you back to the principles you believe in during moments when you have to choose between the right thing and the easy thing.

A vision statement should always include three components:

- Who you are
- What you do
- Where you need to go

The more focused your vision, the more alignment you can expect to see from your organization. Having a vision ensures that your company will make decisions based on long-term goals, even at the cost of a short-term win — and staying on track for long-term goals is critical for any tech business.

If you don't have a vision statement or think that your vision statement isn't serving your organization well, it's time to build one or change the one you have! I suggest that you gather your executives first to discuss and debate what they think the focus of the company should be. Don't be afraid of a little chaos. Instead, embrace the messiness of the process. Have each stakeholder answer each of the aforementioned components of the vision statement. Ask them, "Who are we? What do we do? Where are we headed?" Then share the answers as a group. You may find harmony, or you may find that each executive has a different idea (most likely focused on their sector or area of expertise). Form an amalgam from all these ideas. After you have a rough draft, engage the entire population of the organization. What do people think? How do the answers of the folks in sales differ from those of engineering? Beyond helping to form a vision, this exercise will highlight the challenges in your organization and the areas needing the most alignment.

VISION STATEMENTS FROM FAMOUS BRANDS

The global impact of software enables technology companies to have some of the most ambitious vision statements in the world. Here are a few inspirational visions by companies you're familiar with to inspire you on your journey to build a DevOps culture and create a vision to unify your team:

- **Microsoft:** "Our mission is to empower every person and every organization on the planet to achieve more."
- **Google:** "Our mission is to organize the world's information and make it universally accessible and useful."
- **Amazon:** "Our vision is to be earth's most customer-centric company; to build a place where people can come to find and discover anything they might want to buy online."
- **PayPal:** "To build the Web's most convenient, secure, cost-effective payment solution."
- **BBC:** "To enrich people's lives with programmes and services that inform, educate, and entertain."
- **Whole Foods Market:** "Our deepest purpose as an organization is helping support the health, well-being, and healing of both people — customers, Team Members, and business organizations in general — and the planet."

Ideally, the vision should be integrative — prioritizing people in the organization (and customers outside the company) as well as the technology and product itself. Excellence in both areas is essential to forming a balanced and focused vision, which you need before you can engage in a DevOps transformation.

Incentivizing Your Values

Values are meaningless if you don't incentivize the behavior that lives up to them. Worse yet is to incentivize behavior that goes counter to your organization's vision. So your top priority after creating your vision is to communicate it to the wider organization — and not just the *what* of your vision statement, but the *why*. Communicating your vision is a perfect opportunity to gather the entire team together and get everyone excited about the direction of the company.

SOUTHWEST AIRLINES: A COMPANY WITH STRONG VALUES

I'm loyal to Southwest Airlines for the same reasons most of their frequent flyers are: Southwest is the happiest way to fly. (In fact, I wrote this sidebar while on a flight!) I associate Southwest with smiles and happy people. I used to be terrified of flying, and Southwest flight attendants were always the most patient with me. They offered me water, told me everything was going to be okay, poured me generous amounts of vodka. Southwest doesn't have the best frequent flyer benefits, but I fly with them because they make me feel safe, cared for, and happy. Who doesn't want that?

These values didn't come about by happenstance. It's not as if Southwest just lucked into hiring fabulously happy flight attendants. Instead, the company developed a set of values, built a culture around those values, and then communicated those values to their employees. That is strong company culture.

Your second priority is to ensure that the behavior you want to see from your team is rewarded. I'm a big fan of positive reinforcement because negative reinforcement can be permanently damaging to morale. The focus here is on what you can to do incentivize your values, not how to drag your employees to the proverbial principal's office when they're in trouble. You want employees focused on pursuing excellence, not simply avoiding certain behaviors.

Evaluations

Depending on your company, evaluations can be a time of healthy feedback and personal reflection or a chaotic and panic-stricken period designed to instill fear and dread. You should aim for the former, of course. (If you disagree with my last statement, you should put down this book. I don't want you quoting me.)

Your organization's evaluation rubric must reflect the uniqueness of your company. But I encourage you to include, at the very least, two sections:

- **Team Impact:** This concept refers to the greater impact of the team as a whole. You want to consider the outward impact, such as an increase in the number of users of a service or application, or the launch of a new feature that increased revenue by 10 percent year over year. You also want to consider the internal impact that speaks to DevOps values. This impact includes improved collaboration, better teamwork and communication, reduced silos, and so on. Some of this aspect of the evaluation is difficult to measure empirically or prove causally. That's okay. The key here is that the team gets a grade as a whole, which encourages the team members to evaluate their performance and improve their impact together.

- **Individual Output:** The output of an individual contributor is the summary of their activities. This summary could include features developed, bugs fixed, infrastructure improved, uptime increased, and more. An engineer's output is closely tied to their role within the greater team.

As you may know, I work in Developer Relations (DevRel, for short) for Microsoft. The exact meaning of DevRel varies from organization to organization, but it generally comprises a group of software engineers (or operations specialists, SREs, and others) who sit somewhere between marketing and engineering. It's not a sales engineering role, and those of us in DevRel are never incentivized by sales. Instead, we sit above the sales funnel, gain goodwill for the company we work for within the community, and reflect the wishes of the community back to the product team, ensuring that our applications and tools are as close to what customers want as we can possibly get.

As you might imagine, DevRel is extremely difficult to measure for efficacy. I've settled on this dual evaluation of team impact and individual output, and I think it works really well for DevOps cultures as well. Many of the aspects you love about DevOps, and want to encourage on your teams, are really difficult to measure — especially when evaluating individuals.

Rewards

You may be tempted to throw money at your employees who perform well in your new DevOps culture, and fair market salaries are an absolute requirement. But often, money isn't the best motivator. I know that this idea is a bit counterintuitive. Everyone loves money, right?

Well, yes, to a point. In 2010, Timothy Judge and colleagues published a paper titled "The relationship between pay and job satisfaction: A meta-analysis of the literature" (`https://www.sciencedirect.com/science/article/abs/pii/S0001879110000722?via%3Dihub`). The authors looked at 120 years of research from 92 studies. The results found a rather weak association between salary and job satisfaction. You can read more in the paper published by Tomas Chamorro-Premuzic, "Does Money Really Affect Motivation? A Review of the Research" (`https://hbr.org/2013/04/does-money-really-affect-motiv`), which delves further into the research.

In addition, a 2011 Gallup poll (`https://news.gallup.com/poll/150383/majority-american-workers-not-engaged-jobs.aspx`) found that pay didn't have a significant impact on employee engagement. Essentially, you need to ensure that your salaries are equitable and (if you want to retain your best talent) on the middle to high end of market value. Also, because DevOps values diversity, you should analyze your company's salaries and make sure that everyone is fairly paid. You should have no significant differences in pay among men, women, and people of color. The same work should receive the same pay.

If salary or financial incentives aren't sufficient rewards, your best bet is to reward performance around DevOps and company values by getting a little creative. Following are some ideas I've found to be successful. Remember that your employees and colleagues love to be recognized and appreciated. The more frequently you can highlight performance through a small reward, the happier your engineers will be in the long term.

- **Idea prizes:** Give engineers on both the development and operations sides the opportunity to propose new ideas. The company or team then votes for the best idea and, if the executives agree, gives the person who suggested the idea a small reward. The reward can be a gift card to their favorite coffee place or tickets to a baseball game. Honestly, the reward doesn't even have to be worth anything monetarily. I've seen companies reward ideas with coveted stickers and LEGO pieces that engineers can then show off on their machines and desks.
- **Hack time:** Listen, engineers love to do just that — engineer. You'll earn their loyalty if you give them dedicated time to work on the side projects that excite them. When the team accomplishes something, give them a week or two to work exclusively on something of their choice. It could be an open source project, an idea for something to improve their everyday work, or something altogether unrelated to your business. If the hackathon produces something usable by the company, it's a bonus. The purpose is to give your engineers paid time to work on passion projects.
- **Fun off-sites:** Sometimes the team needs to step away from the office and engage in a different activity to build rapport and trust. No, this isn't your average foray into a ropes course or trust falls. Don't try to design trust; it doesn't work that way. Instead, give your team the opportunity to get to know each other in a more informal way. The activity needs to be inclusive so that everyone can participate, but beyond that, anything goes! I love bowling because it can be silly and I'm terrible at it. But you could volunteer, take dance lessons, go to a yoga class, or take a trip to the mountains. The specific activity means much less than the opportunity to have fun together away from the office.

Inclusive off-sites can be a blast but require a bit of consideration. Ensure that you're providing your employees with what they need to travel. At a basic level, provide every employee their own hotel room, transportation, and food. After you have the basic issues managed, consider the needs of each individual. Single parents might need financial or logistical help to find childcare. Breastfeeding mothers may need you to pay for shipping breast milk home on dry ice. Newly sober alcoholics may need help staying away from alcohol. Employees with disabilities will need environments that work for them. Paying attention to the tiny details is what will help everyone relax and have fun.

» **Removing the seven types of waste**

» **Getting to market faster than your competition**

Chapter 3

Identifying Waste

After you have a clear idea of what your DevOps culture will look like, it's time to review your current processes and look to the future for improvement. DevOps has three focuses: people, process, and technology. Process is second only to culture in a DevOps transformation.

Process is the area in which you'll see the most quantitative improvement in the speed of your organization's software delivery. But this chapter doesn't focus on how you improve your processes. (I discuss making process improvements to every phase of your software delivery life cycle in Part 2 of this book.) For now, think about your team's software development processes holistically. Step back and see it as an ecosystem of people implementing processes with technology.

Waste is any activity that does not directly impact the experience of the customer. If an action, activity, or process doesn't add value to your customers, it's wasteful. Increasing your team's velocity with DevOps requires you to identify and eliminate waste.

You would be shocked at how much waste you have in your development process. In fact, the Lean Enterprise Research Centre (LERC) at Cardiff University in the U.K. has found that up to 60 percent of the activities that engineers routinely engage in are wasteful and have zero impact on the end user. That's . . . disturbing.

Understanding the different categories of waste helps you identify the easily improved processes of your system. Think of this initial list as the low-hanging fruit by which you can see quick wins in your DevOps transformation. The faster you can apply the benefits of DevOps, the more smoothly your transformation will go.

In this chapter, you discover the seven categories of waste in complex systems, learn how to collect data and identify bottlenecks, and prioritize the customer by focusing on impact.

Digging into the Seven Types of Waste

I believe that the average farmer puts to a really useful purpose only about 5% of the energy he expends . . . . Not only is everything done by hand, but seldom is a thought given to a logical arrangement. A farmer doing his chores will walk up and down a rickety ladder a dozen times. He will carry water for years instead of putting in a few lengths of pipe. His whole idea, when there is extra work to do, is to hire extra men. He thinks of putting money into improvements as an expense . . . . It is waste motion — waste effort — that makes farm prices high and profits low.

— HENRY FORD, *MY LIFE AND WORK* (1922)

Many DevOps principles are rooted in lean manufacturing, a principle that emphasizes identifying and eliminating waste in order to improve production velocity. Lean manufacturing identifies seven types of waste. I've ordered the types of waste in this section by most-to-least impactful. In other words, the first type of waste listed is likely your lowest-hanging fruit and the one that you should tackle first.

Unnecessary process

Process is a huge component of DevOps because it streamlines activity, behavior, and expectations in every aspect of your business. But process can quickly become an enemy. How many meetings are your engineers required to be in every week? Do your daily standups take fewer than ten minutes, or does the time spent make sitting down necessary? Another insidious cause of unnecessary process takes place when product requirements aren't clarified at the beginning and work has to be, well, reworked.

Waiting

Inaction at any part of the development life cycle — the time from when you plan to develop a piece of software to the time you deploy it — is waste. Yet, your organization is probably riddled with waiting. Engineers wait for QA to test a new code. Infrastructure waits for developers to build products for deployment. Developers wait for infrastructure to provision new machines. Everyone waits for everyone else to supply siloed information. Waiting is common and difficult to combat.

Motion

Think of motion as busy work. It's the wasted activity that you and your team complete. In site reliability engineering (SRE), this work is referred to as toil. If an activity doesn't have impact on your customers, its purpose could be to "look" good. This work could also be the result of inefficient processes. The former, "looking good," relates to your incentives and review processes. The latter points to where automation can begin to speed up your team's efficiency in a major way. Technology itself can also produce the waste of motion. Perhaps you're paying for infrastructure or tools you don't use.

Costs of defects

Defects are one of the most easily recognized types of waste. In car manufacturing, one type of waste might be scrap metal. In software, defect waste includes bugs and technical debt. You should also include service downtime in this category. Anytime an engineer has to "fix" completed work, you're in defect territory. I'm not a big fan of the "just engineer better" approach because there will always be unknowns and edge-case bugs. Your ability to combat this waste will be in your team's forethought in architecture to ensure expected behavior and responsiveness to quick iterations. You want to ensure that the "blast zone" — that is, the customers and services impacted — is small and that every engineer has the ability to respond to bugs easily. (See more about responding to bugs in Chapter 17.)

Overproduction

In manufacturing, overproduction refers to any excess parts or products produced that the company can't use or the customer is unwilling to purchase. In software, overproduction comes in two forms: wasteful code and products that don't meet the market's needs. You want to avoid having software developers work on solving problems that don't exist or overengineering the solutions. But you also want to make sure that the products you produce and bring to market are desired by the customers you're trying to reach.

Transportation

Transportation waste takes place anytime a product, person, or tool is moved from one location to another. Now, unlike Toyota, for example, you don't have to ship cars from the assembly plant to dealers across the country. But you do move code between servers and repositories. You also move people between teams, which requires time to adjust and get up to speed.

Inventory

Most likely, inventory is much less of a challenge for you and your company than, say, a car manufacturer. Few companies ship physical software these days, and inventory has become less of a problem. Still, you can have inventory, and any valuable product that is waiting to be sold or used is wasteful. Think about something as simple as the five laptops you have sitting in a room somewhere in the office because you've had engineer turnover and are waiting to hire new employees. You could also eschew the concept of physical inventory and consider code and proprietary information to be your inventory when evaluating waste in your organization.

Understanding waste in DevOps

Waste comes in many forms. No two pieces of waste will be the same, and your approach to eliminating waste will need to adapt to new challenges. In fact, after you get started tackling waste in your software delivery life cycle, you discover that it's occasionally like playing the game of whack-a-mole: You eliminate one piece of waste only to see another pop up later in the life cycle as a result of your change.

DevOps takes several of its core ideas from lean manufacturing, a management philosophy distilled from the Toyota Production System (see the "Principles of the Toyota Production System" sidebar for more information). Lean manufacturing uses three separate Japanese words to describe waste:

- *Muda:* Waste
- *Muri:* Overburden
- *Mura:* Unevenness (or irregularity)

Start considering how you would approach *muri* versus *mura.* Where do you see these three definitions of waste in your current processes? Do you have employees who have been overburdened to the point of burnout? Are all your engineers carrying the weight of your workload evenly or do you have extremely high and low performers? How might these definitions of waste apply to all three areas of DevOps — that is, people, process, and technology?

REMEMBER

An important point to keep in mind when tackling waste is to improve efficiency through optimization and simplification. But also remember that waste almost never originates from a place of bad intentions. In fact, most waste exists because of inertia. Habit is the worst enemy of efficiency in an engineering organization. "We've always done it this way" is poison that rots fresh ideas at the root. Do your best to eliminate that phrase from the minds of everyone in your organization. When you eliminate waste, you improve quality, reduce development time, and lower costs.

PRINCIPLES OF THE TOYOTA PRODUCTION SYSTEM

Originally referred to as "just-in-time production," the Toyota Production System (TPS) was built on the manufacturing philosophy of Toyota founder Sakichi Toyoda. The TPS business philosophy predated lean manufacturing and emphasized continuous improvement and eliminating waste.

The TPS management approach is detailed in the book *The Toyota Way* and breaks the system into 14 principles, all of which can buttress your DevOps practice:

- Emphasize long-term reputation, even at the expense of short-term financial losses.
- Reveal problem areas by creating a continuous process flow.
- Focus on your key value-add and avoid the overproduction caused by executing every "good" idea.
- Don't burn out people or overburden equipment.
- Prioritize quality and empower everyone to stop the process when necessary.
- Standardize processes to provide consistency.
- Create visual tools for everything so that problems can't be hidden.
- Put technology second to people and processes.
- Train and educate employees.
- Grow employees who believe in the company's culture and philosophy.
- Help business partners improve.
- Managers must "go and see" the work first-hand so that they understand the challenges of their engineers.
- Decide slowly and implement decisions quickly.
- Reflect (*hensei*) on feedback and continuously improve (*kaizen*) to serve the customer.

Successful tech companies understand their customer's pain points and respond to those needs through well-designed products. Continuously improving quality is what separates those organizations from others that fizzle out (or burst into flames). Software delivery takes time, but if you can reduce your time to market, you reduce engineering costs and increase the likelihood of capturing more market share. Reaching customers as quickly as possible provides the opportunity for feedback and iteration.

Each of the wastes identified in lean manufacturing have associated costs. Tackling even one will significantly impact your organization's bottom line and allow you to reduce total costs.

Rooting Out Waste

How do you go about identifying waste, simplifying your process, and reducing costs? Well, you could play pin the tail on the waste donkey and just pick an area of waste to focus on. Or (and this is the path I personally recommend) you can be more purposeful in observing your software development life cycle holistically and identify the most impactful areas to mitigate first.

Making sweeping changes and measuring your success are impossible without knowing where you started, especially if you need to coax executive buy-in for your DevOps transformation. Here are the three types of actions to identify within your software development processes:

- Wasted actions to be eliminated
- Wasted actions that are necessary within the current system
- Actions that add value to the process

Observing will be the best use of your time at this stage. Start with people. For example, are the meetings that engineers attend wise uses of time or pointless *motion*? Next, look at process. Does a manager have to sign off on releases before a developer can deploy code to production? Could that be *unnecessary process* and *waiting*? Finally, observe your tooling. How many bugs make it into production? What are your *costs of defects*?

Discovering bottlenecks

One of the most insidious forms of waste is a bottleneck. The term *bottleneck* refers to a congestion or blockage along a process. Just as a bottle narrows at the neck, so too can processes. Imagine a wide river that is capable of allowing a dozen boats to sail in parallel. If at some point the river narrows (illustrated in Figure 3-1), the boats will have to sail one at a time, creating congestion (shown in Figure 3-2). This narrowing slows (or sometimes halts) production. Ideally, you identify the bottlenecks in your own processes and enable engineers to use DevOps to make the proverbial river wider and allow for more work to flow at the same time.

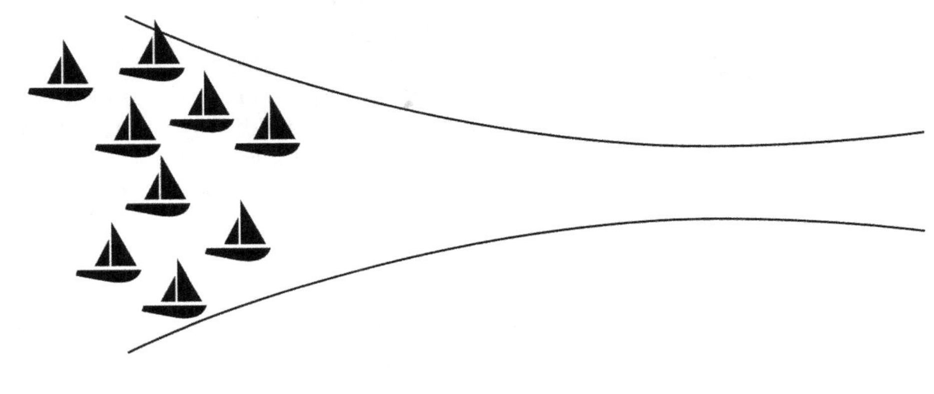

FIGURE 3-1: A wide river is about to narrow.

FIGURE 3-2: A bottleneck significantly slows the flow.

Bottlenecks can occur at any point in a process. Two of the most common bottlenecks I see come in the form of approval processes and manual tasks. These bottlenecks can result from mandated manager approval before releases, or reliance on a manual deployment process that is owned by one person (who gets busy and occasionally goes on vacation).

Congestion can also occur when you fail to address concerns early on in the software delivery life cycle. If you wait until you release code into a staging environment to confirm that the code is secure, you'll likely have to kick the code all the way back to the development phase. Addressing security concerns in the planning process can avoid wasting time and engineering resources.

PANAMA CANAL: THE WORLD'S LARGEST BOTTLENECK

The Panama Canal was completed in 1914 and is possibly the most important waterway on the globe. Approximately 5 percent of all trade flows through the canal. On average, 34 ships go through the canal every day, and each ship requires 52 million gallons of water to move through the waterway. Ships have evolved significantly since the canal was built and, unsurprisingly, those ships have become a lot bigger. A few years ago, the canal underwent major construction to double capacity and accommodate the latest generation of enormous container ships. Prior to the expansion, long lines of ships waited to pass through. The queues reached such lengths that *Disney Magic,* a cruise ship, paid more than $300,000 just to jump the line. If the Panama Canal Authority hadn't addressed this bottleneck, it would have lost out to its competitors, the Suez Canal and U.S. railways.

After you start looking for bottlenecks, you might feel overwhelmed by just how many exist in your current system.

Tech companies experience two forms of bottlenecks:

- **Short-term bottleneck:** Caused by a temporary hiccup. Your most reliable engineer goes on vacation, for example.
- **Long-term bottleneck:** Results from consistent, compounding friction in the production process, such as a slow machine that results in a long queue of inventory.

The cause of a bottleneck usually comes down to one (or more) of three reasons:

- **Capacity limits:** The machine or tool has reached its highest capacity. It needs to be replaced or improved, or additional resources need to be added to the system. Sometimes a bottleneck occurs because a team has few engineers. This lack is particularly visible when a team's skill sets are unbalanced, such as when one or two operations engineers are supporting the work of dozens of developers.
- **Inefficient use:** The resource is not fully utilized. If the bottleneck is caused by a tool or machine, you might have a tuning issue or are perhaps using the wrong technology. In the case of humans, you could be underutilizing someone's talent by pigeonholing them into a specific role when they would excel elsewhere.

» **Underqualified engineer:** Software engineers are under constant pressure to learn the next greatest technology. Sometimes fixing a bottleneck is as simple as providing the necessary training and continuing education. (I discuss empowering engineers in Chapter 15.)

Anytime a bottleneck occurs — whether from wait times, overloaded machines, or exhausted humans — it stalls production. In other words, that one bottleneck slows the entire production chain and creates a queue of units that need to be processed. The situation is not exactly fun when a bunch of executives are breathing down your neck.

When you've identified your bottlenecks, evaluate the degree of impact. Major bottlenecks should be addressed as soon as possible, whereas minor bottlenecks are much less concerning.

WARNING

Perfection in your production flow is impossible. If you chase perfection, you'll spend more time trying to locate every single bottleneck instead of removing the waste that's causing the biggest problems. Don't worry about each little thing. Instead, focus on the one or two things that have the biggest impact on your development process.

Focusing on impact

One of the best ways to reduce waste and eliminate bottlenecks in your development cycle is to focus on impact, which you do by prioritizing the work that has direct impact on your customers. If something doesn't matter to your users, it shouldn't matter to you. (Or it shouldn't matter much!)

When faced with solving for a bottleneck, you have a couple of options, as described in the following sections.

Increase your number of employees

Adding head count to your organization can seem like an easy fix for a bottleneck situation, and sometimes it's just what you need. People-centered bottlenecks left untreated are like poison to teams. Your engineers burn out and morale across the team suffers. Adding fresh contributors (and new ideas) to the team can breathe new life into your engineering processes. Here are the pros and cons of increasing your head count:

» **Pros:** Human redundancy helps significantly with responding to increased demand as well as with managing employee vacation time, unexpected illnesses, and planned family leave.

» **Cons:** Having more cooks in the kitchen can increase communication complexity and requires time to bring those new employees up to speed. Hiring and training takes both time and money.

Eliminate unnecessary activities

If an activity doesn't add value, cut it. I guarantee that your team completes work daily that is almost entirely unnecessary. Such pointless jobs often stem from "the way we've always done it" or a lack of automation. If a redundant task is manually completed, automate it. Occasionally you'll discover that you can completely remove the activity from your process with no impact. Here are some issues that arise when you start eliminating unnecessary activities:

» **Pros:** Eliminating unnecessary activities is one of the easiest steps you can take to reduce waste. Just give your engineers permission to stop doing work that doesn't matter. If an activity turns out to matter after all, you can always add it back.

» **Cons:** Make sure that you understand the problem and the solution before you automate a fix. The wrong solution can create a problem that's much, much worse than the waste it was meant to fix.

Provide a buffer

Make your team asynchronous. That is, if a single point in your development cycle requires waiting, put enough buffer work in place for the engineer or team to be doing *something* while they wait. For example, you should have a backlog of engineering work that needs to be completed, but not urgently. Often this backlog will include technical debt — work that was deprioritized or deferred in order to make deadlines. (Technical debt includes work like refactoring a poorly implemented function, adding tests to ensure functionality and consistent performance, and creating shared libraries to eliminate duplicate functionality.) Another option for creating a bottleneck buffer is to encourage engineers to learn new skills or experiment with new technology while they're waiting. Here are issues to consider when providing engineers with work to do while waiting:

» **Pros:** If you can't remove a bottleneck, having a buffer is a good solution to make the entire production system work together. You still want to try to eliminate the bottleneck at some point, but the buffer buys you some time.

- **Cons:** Context switching can absolutely crush productivity, and although a quick fix can be addressed while waiting, this is not the time to throw extremely complicated problems at your engineers. Make sure to break extra tasks into manageable pieces.

Ultimately, the best way to prevent bottlenecks is to train your engineers on every aspect of the process. No, I don't expect developers to be experts in Kubernetes. I also don't expect operations folks to be pumping out features in Java every week. But cross-training provides a certain level of adaptability that enables your engineers to find workarounds and reduce downtime. It also reduces confusion when work is handed off from one team to another.

» Persuading your peers

» Gaining executive buy-in

» Responding to pushback

Chapter 4
Persuading Colleagues to Try DevOps

When I'm on the road a lot, talking to engineers, they often ask where they should start. "DevOps sounds great," they say, "but what's the first step?" or, "My boss has decided we should 'do the DevOps,' and has reorganized us into a DevOps team, but what are we supposed to be doing?"

Everyone's DevOps journey is different —unique to you as an individual, to your team, and to your company. You will pick and choose (to a certain extent) which aspects of DevOps will benefit you most and apply those aspects to your team. One thing is certain, however: You can't go it alone. Your DevOps transformation will fail if you attempt to force the new way of thinking onto your team without first persuading them. You must sell your colleagues on the benefits of DevOps and energize the organization around new possibilities.

In this chapter, you dig into why humans loathe change, work on perfecting the art of persuasion in order to effect change, practice explaining DevOps to leadership, and see how to respond to doubting minds.

Fearing Change

Humans don't like change, and the reason is based in our brains. Habits are powerful because they're efficient. Your brain can think less and still achieve the same amount of productivity. Your brain is extremely proficient at processing information.

Psychology offers information on why people resist change so strongly. Inertia is powerful and change is expensive. People are likely to stay on the path they're already on because shifting that path takes quite a bit of effort. Staying the course is much easier. When you do decide to climb your way out of your current groove, persisting at it takes an extraordinary amount of brain energy. (Ever notice that you're a little more hungry when you're learning something new?)

In addition to the inertia aspect of change resistance, two other key aspects make people fear change. Keep both these things in mind as you go through this chapter:

- **Past experience:** Every single person in your organization comes to their job with years of history that have chalked up successes, failures, and fears. Some people within your company have likely watched changes made in their past workplaces succumb to failure. Failure stings. Some of your colleagues may have even lost their job over a massive failure. Fear of repeating such experiences doesn't just disappear.
- **Uncertainty:** Your brain is more likely to categorize uncertainty as a threat rather than an opportunity. Evolutionarily, this tendency was important to keep humans, well, alive, and that tendency persists even though most of us aren't chased by lions these days. Also, change usually doesn't happen overnight, which forces people to take a wait-and-see approach although their brain desires to know the outcome now. This situation creates conflict. Sometimes the conflict is internal as someone weighs their fear of failure against new possibilities for success. Other times, the conflict surfaces between people. You may adapt to change more quickly than your colleague, and that delta in time required to transition can introduce interpersonal friction.

Despite the natural fear of change, the capacity for change is critical to the survival of any business. Examples abound of businesses whose internal resistance to change sealed their eventual fate. To cite just one example: Remember Blockbuster? (My family had a Friday night tradition of hopping in the car and heading toward the royal-blue sign down the road. Each of us would spread out over the store, pick our individual favorite, and then have it out over which one or two we should rent.) At its peak, Blockbuster had nearly 10,000 stores. In 2000, Netflix offered

Blockbuster a deal to acquire Netflix for $50 million. The Blockbuster CEO declined; Blockbuster wasn't interested in the "niche business." You know how that story ended.

The leaders of Blockbuster weren't idiots — far from it. But they, and many others, failed to see the writing on the wall and correctly predict where the market was headed. They also failed to communicate with customers and, ultimately, failed to change their business to conform to what the market wanted.

Persuading Those around You to Shift to DevOps

Empathy is a powerful tool, and showing true understanding of the fears and doubts that people around you experience can help your DevOps transformation succeed. Simply acknowledging the potential fears of your colleagues can go a long way toward assuaging their anxiety and persuading them to get excited about the new possibilities that DevOps provides.

One way of working with the natural human resistance to change is, first, to understand and expect it (see the preceding section, "Fearing Change") and then to hone your skills at persuasion. I like to think of persuasion as tailored messaging. It's presenting an idea in a way your audience can understand. That doesn't mean that you have to build separate arguments and pitches for every person with whom you come in contact. Instead, keep in mind the four most common styles of leadership that people embody in problem-solving. Basing these styles on the Myers-Briggs personality types, you can group people as visionaries, strategists, administrators, or counselors. (Obviously, these categories oversimplify people, but they enable you to ensure that your arguments for DevOps persuade even the most stubborn.) Keep in mind the four personality types when you talk about DevOps to your executives, peers, employees, and business stakeholders. Each of these personality types will relate best to the following approaches:

- **Hope and imagination for visionaries:** The thinkers are intellectually curious more than anything else. They want to see the data. But they also want to hear about a world of tech that doesn't exist yet — a world that they have the chance to build themselves. How has DevOps improved the processes at other companies? What are the big advantages? After you've given them enough information to get over their initial hesitation, you can think of them as mental petri dishes. All you have to do is prime them and they'll inquisitively dig in further — growing your argument for you.

- **A high-level plan for strategists:** You don't need to dig into the details for these people. These quick-witted folks are creative problem-solvers. They are also your risk-takers, which makes them some of the easiest to persuade. They'll find the change to DevOps exciting, and their natural curiosity will energize them to your side. Just be sure to have room for them to contribute. They'll likely want to know how the transformation is progressing and how they can be called on to energetically persuade others.
- **Detailed direction for administrators:** The worker bees keep the hive buzzing. These people do the work diligently and will be responsible for carrying out the strategy set before them. They are meticulous, dependable, and organized. Use the fact that DevOps is an incredibly practical way to ensure that the system runs smoothly, from determining requirements to shipping software.
- **People-centric pathos for the counselor types:** Pathos — emotion — will be the most effective persuasion tool for people who tend toward being caregivers. They put people first, no matter what. Understanding how DevOps helps to smoothe communication, reduce interpersonal friction, and increase collaboration will soothe this group's fears.

In addition to knowing how to approach the various personality types, it helps to have a clear sense of the three main groups within your company that you'll need to win over to the DevOps philosophy: executives, managers, and engineers. Figure 4-1 represents these three groups. The hourglass shape with managers in the middle isn't meant to suggest that the managers in your organization aren't important to your mission; quite the opposite: They're critical to full adoption. But they can be the most difficult group to persuade, so I suggest that you tackle them last. (They're the last to come out of the hourglass either way you turn it.)

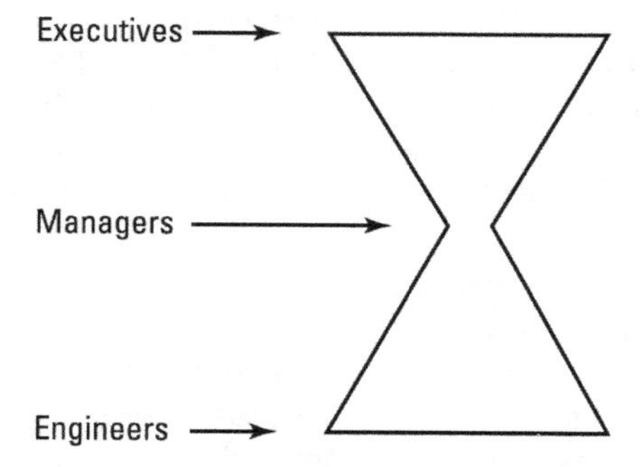

FIGURE 4-1: Persuading each group in your organization.

The reason I focus on executives and engineers first is that managers sit between a rock and a hard place, constantly facing scrutiny from executives and mutiny from engineers. As a result, they are naturally conservative decision-makers. They are comfortable within the status quo because they know that any change will ripple through them and cause strife somewhere along the chain of command. If you can get buy-in from executives and support from engineers, managers will have zero reason to protest. Finally, if engineers float DevOps to their managers and expect the manager to relay the message to executives, the purity and passion of the argument is easily lost. Direct contact between engineers and executives prevents miscommunication and unnecessary friction introduced by fearful managers.

For executives and engineers, you have a choice of which group to approach first. If you're effective, either will provide a great groundswell of excitement. Executives will provide clout and affirmation around your vision. Engineers will provide a massive amount of people who are more than willing to explain why they need smoother development processes to produce better software faster. Just be careful to focus your energy.

IDENTIFY EVANGELISTS

One of the keys for getting a DevOps mindset under way in your organization is to identify evangelists. You can't transform a company alone, and you certainly won't accomplish a full-fledged culture change by yourself. You need others to believe in the mission you're working toward and help you spread the message.

Beyond the obvious benefits of creating excitement and earning buy-in from your colleagues, building a small team of evangelists hedges you against burnout. Leading your team to the DevOps promise land is quite the journey. It's long, exhausting, and full of landmines. Maintaining your excitement and passion is critical to your success. Surrounding yourself with people equally motivated toward a DevOps transformation will keep you going.

Evangelists are people whose influence ripples out among the rest of the team. They act as multipliers, and igniting one evangelist will earn you the support of many rather than one. Locating potential evangelists on your team is a better way to focus your time early on. If you face too much pushback too quickly, you risk burning out and giving up. Look for evangelists who are different from you and can communicate well with those with whom you may struggle to find common ground. For example, if you're a front-end engineer, look for an evangelist on the operations side who can talk about aspects of DevOps that you can't.

Earning executive support

Of the main groups that you need to get on board with DevOps, executives may be the most important to your cause. A DevOps transformation is nearly impossible without their buy-in. Other groups can subvert your efforts and quietly create friction, but executives are the only group that can squash the project altogether — and in just a few sentences.

In Chapter 3, I talk about the various types of waste in engineering teams and about identifying bottlenecks along your process. This information is crucial to executive support for your DevOps transformation. Executives often focus on vision, the big picture, but they also love data. You can hook them with your enthusiasm and then finish selling them with data, analysis, and a plan. As Brené Brown says, "Maybe stories are just data with a soul."

Gaining the support of your executive leadership is a big win. You will be Sisyphus without it. Executive leadership gives you key advantages that will help the transformation process go more smoothly. They control budgets and team head count (the number of people allocated to a project). They can also lend quick fixes to conflict. Also, if you managed to convince one or two executives that DevOps is a worthwhile cause, they will help you persuade the others from inside the boardroom.

You need more than vision to convince these folks. You also need to tap into their dreams for the company, as well as their fears. Think of the pressure your executives are under from a public perspective. Your CEO can't be the one to lose the company. Your CTO can't afford to lose out to your competitors. You can acknowledge those fears and use them to tap into executives' emotion and hook them. Then you can provide the supporting evidence.

Every year, DevOps Research and Assessment (DORA) releases the *State of DevOps Report* (`https://cloudplatformonline.com/2018-state-of-devops.html`). It provides diligently collected and analyzed data from the tech industry and provides rich data for you to use in your argument for DevOps. According to the report, elite performers — companies who deploy on demand and generally recover from incidents in under an hour — outperform companies that have low DevOps adoption by significant amounts.

In 2018, Elite DevOps organizations

- Deployed code 46 times more frequently
- Had a 2,555 times faster lead time from commit to deploy to production
- Recovered from incidents 2,604 times faster

Creating a groundswell in the engineering group

I like the word *groundswell* because it embodies the imagery to think about as you advocate for DevOps within your organization. A *groundswell* is a series of tightly grouped waves — adored by surfers — that last more than 15 seconds and are caused by a storm thousands of miles away.

Imagine sitting on a calm beach and then seeing a slow momentum building in the water, eventually rushing toward shore. It's unstoppable. That is the power your team of engineers will give you if they adopt your view of DevOps and buy into the culture, the philosophy, and the approach. I use three tactics when approaching (sometimes doubtful) engineers about a DevOps transformation:

- **Ask questions.** What are your engineers struggling with right now? Find out where their pain points are, and then discuss how DevOps may be able to address them. If they're already doing well with code management, releases, and production deploys, talking about continuous integration and continuous delivery won't get you the traction you need. Instead, talk about how annoying it is for developers to go through an operations person to get certain log files and application performance data. Or how frustrating it is that a handful of ops folks are on call and engineers don't contribute to maintaining the applications and services they build. (See Chapter 19 for how those issues are handled, or don't even occur, in a DevOps system.)
- **Offer concrete suggestions.** Engineers like evidence. They also like to see you've thought about how to address issues before talking about them. If you go to the engineers with a bunch of lofty ideas and no execution strategy, the conversation might not go the way you expect it. Instead, think through which challenges you want to tackle first. If you've identified waste in your development process (explained in Chapter 3), you have a good idea of where the low-hanging fruit is. If you haven't taken the time to look at potential bottlenecks, take a good guess and come up with a few DevOps approaches to improving the current situation.
- **Encourage your engineers to experiment.** The best part of persuading your company to adopt DevOps is that you don't always have to do it with words. Instead, you can simply start practicing the philosophy and approach. Allowing engineers to experiment through small projects allows them to experience the visible difference firsthand. Sometimes it's best to beg for forgiveness instead of asking for permission. Just do it. Keep it small and be sure to brag about how awesome your little experiment went.

Managing the middle managers

Middle managers often comprise the most difficult group of people to convince that DevOps is a smart approach to software development. They were promoted to their current position because of their prior work, so getting them interested in shifting directions from the path that has brought them success can be a tall order.

Kodak serves as a great example of this challenge. Before it became a dinosaur, Kodak was a highly inventive company. It consistently adopted new technologies quickly, including digital technologies in photography. Part of Kodak's problem, though, was that its advances were too spread out through the market. People — even employees — simply couldn't see how innovative Kodak was because the company's small, yet impressive, innovations were hidden in a vast web of products. The company lacked focus and an organized strategy.

When George Fisher came on as CEO at Kodak, he moved everything into a single division whose sole purpose was to launch new products. Internally, Fisher faced pushback for his "aggressive" strategy. Middle management never got on board. They fundamentally didn't understand that the industry was shifting and Kodak was quickly losing market share. The situation was urgent and needed quick action to mitigate. Yet the Kodak middle managers felt threatened by the changes, and their resistance was one of the last nails in Kodak's coffin.

Middle managers matter. A lot. They're the individuals who will pass the vision of the executives down to the engineers who are closest to the keyboard. They're also the intermediaries who help executives understand what is and isn't possible from an engineering perspective, so getting them on board is important. Still, I suggest that you work on persuading this group last. The process of convincing them will flow much more smoothly if you take advantage of the peer pressure from the other groups. Get executives and engineers excited about the potential problem-solving that DevOps brings to your organization and *then* capture the attention of managers. After you have everyone else on board, it will be an easy sell.

Persuading the stubborn

So, how do you persuade the executives, engineers, and managers who remain stubbornly resistant? I once read about a sales approach that involved identifying two people in any room you enter. One of those people is your advocate — the person who will pull for you, speak for you, and protect your point of view in meetings to which you won't be invited. The other is the person who is the least impressed with you. That person will quietly argue against your suggestions.

I try to apply that technique, and although I don't always get it right, it's an interesting exercise to try it. Some of your colleagues are likely to be on your side immediately. They'll be thinking, "Wow! Fewer ineffective meetings and stressful deploys, less downtime, more cooperation, and faster development? Where do I sign?"

Others, however, will be extremely slow to come around to your way of thinking. They'll drag their feet and suggest alternatives. They'll wonder how your suggestions are any different from the thousands of approaches they've seen companies adopt before — approaches that have either failed miserably or not produced impressive improvements.

Look at the situation from their point of view. What's the point of putting in all this effort for minimal results? From their perspective, the company might as well keep going in the direction it's currently headed, and keep doing things the way they've always been done. After all, the company's situation is not *that* bad. It deploys good software. It has bugs, sure, but doesn't everyone? The customers are mostly happy. What's the impetus for changing?

Well, you can supply them with all the facts you've learned after you've completed this book.

Or you can choose not to do that. Seriously, at some point you may have to decide to abandon your persuasion efforts and simply get on with implementing changes, adopting new practices, and automating manual tasks. Where that point is, exactly, will depend on you and your company. But you're likely to know when your ideas have gained enough of a foothold to make turning back the tide impossible, such as after you've convinced 70–80 percent of the key influencers in your company. But you'll know. You won't be able to get everyone excited about these ideas, and that reality has nothing to do with your presentation or the merits of DevOps. Some people are simply stuck in their ways, and no amount of groundswell or data will change that fact.

Understanding the Adoption Curve

Sociologists use adoption curves to model how people adopt new innovations. Though adoption curves have been adapted for many industries and purposes over the years, the clusters of adopters were first grouped by agricultural researchers George M. Beal and Joe M. Bohlen in their 1957 paper *The Diffusion Process* on how innovative farming practices are adopted.

The original curve clustered people into five groups: innovators; early adopters; early majority; majority; and non-adopters. The original names were meant to associate the groups with the overall adoption by the larger population. For example, "early majority" refers to the group that sits at the cusp just before the majority of the population has adopted the innovation.

After the work of Beal and Bohlen, Geoffrey Moore popularized the adoption curve for tech by in his book *Crossing the Chasm*. The numbers at the bottom of Figure 4-2 refer to the percentage of the population. Trendsetters and early adopters represent about 30 percent of total adoption, whereas the late majority adopter group represents almost 80 percent adoption.

For Figure 4-2, I've tailored the adoption curve to DevOps to show you how you can expect your colleagues to warm to DevOps as you persuade them. The early innovators, trendsetters, and trailblazers in your organization will dive head first into DevOps without a care in the world. Others will follow them soon after. Later, after you've built some momentum with your new system, you'll see the early and late majority join the club. At that point, you can feel confident that DevOps will embed itself in your company regardless of whether the curmudgeons get on board.

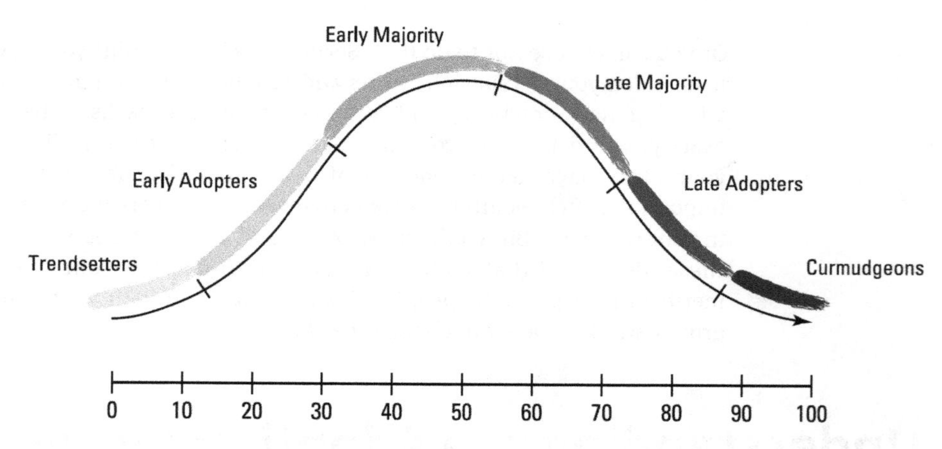

FIGURE 4-2: The DevOps adoption curve.

COMMUNICATING THE MESSAGE

The "people" aspect is the largest and most important part of DevOps. And yes, if you're technical, the emphasis on persuasion skills can feel a bit juvenile — maybe even beneath you. You're not an outlier in feeling like that. In my experience in working with organizations, a lot of engineers have that reaction when they start digging into DevOps.

But my job with this book is to arm you with the tools you need to successfully transform your organization to one that follows DevOps principles, and persuasion is one of the biggest tools you'll need in your arsenal. DevOps is a huge, encompassing philosophy that, like Agile, you can apply in a thousand ways. The application and implementation of DevOps is far less important than the outcomes. Your outcomes depend heavily on your customers, your current culture, and your industry. Your job is to understand all the aspects of DevOps, choose the parts that work well for you, and then put the other pieces to the side. Your choices don't have to be permanent. Nothing is. But those choices will allow you to focus on the pieces of DevOps that are most likely to give you the best outcomes.

When talking to anyone at your company about DevOps, remember that so much of the philosophy is about collaboration. There is no one-size-fits-all approach. There also aren't any must-haves or requirements. So be open to suggestions and allow flexibility in your approach, and you'll be amazed at how much consensus you can build rather effortlessly.

In park management and transportation planning, natural paths formed by erosion from animals or humans walking over the same piece of soil over and over and over are called "desire paths." They form naturally along the paths that are the most efficient.

Communication tends to follow similar patterns to such desire paths. It's difficult to predict who will jell most with others on a team and who will communicate most seamlessly. That knowledge comes with observation over time. Yes, some people seem to be natural communicators, but effective communication is also a learned skill — one that you or anyone can master with practice.

If persuasion isn't your natural talent, don't worry. Remember to identify evangelists (see the "Identify evangelists" sidebar, earlier in this chapter) and know that leading your team in a DevOps transformation doesn't require a natural propensity toward persuasion or oration skills. Most of the conversations you'll have over the first

(continued)

(continued)

part of this transformation will be with one or two people. You're most likely not presenting to large audiences. But you should do two things before you approach a colleague about DevOps:

- **Prepare:** Know whom you're talking to and do your best to predict their concerns or questions. That way, you can be prepared for anything that comes up. If something unexpected happens, simply say, "That's a great question. Let me do some research and get back to you."
- **Practice:** Know what you want to say, how you want to say it, and — most important — the one thing you want your audience to leave thinking. Jot down what you want to say. Write it out word for word, make bullet points, or do whatever else works best for you. It may feel silly, but stand in front of the mirror and imagine talking to one of your colleagues or an executive. Practice helps you feel more confident before you ask people to adopt a new development philosophy.

Pushing for change

Gartner, a technology research firm, created the graphic presentation of what it calls the *hype cycle*, representing the stages of maturity and adoption of specific technologies.

You can see a version of Gartner's hype cycle displayed in Figure 4-3. Though the hype cycle is generally used to describe the public's perception of a technology, I believe that it also applies to what you may feel during the first few months of introducing and implementing DevOps. The cycle has five main phases:

- **Trigger:** The project kicks off. You've made no major changes yet and haven't received feedback.
- **Peak of Inflated Expectations:** You're beginning to talk to some people and have generated excitement. Everyone seems to love the impact that DevOps could have on the team, and you're expecting a relatively frictionless cultural transformation.
- **Trough of Disillusionment:** Curiosity and excitement begin to wane. The reality of implementation complications and failures are beginning to weigh on you, and you're receiving more executive pushback than expected.
- **Slope of Enlightenment:** The realities and hardship of transforming your organization into a DevOps culture are leveling, and the team begins to have a clearer vision of what needs to be done. You've received some excellent feedback and know how to get the team to work together to iterate and improve.

» **Plateau of Productivity:** Production and emotion level off into a steady state of continual improvement. You're on your way and can see small but important changes taking place.

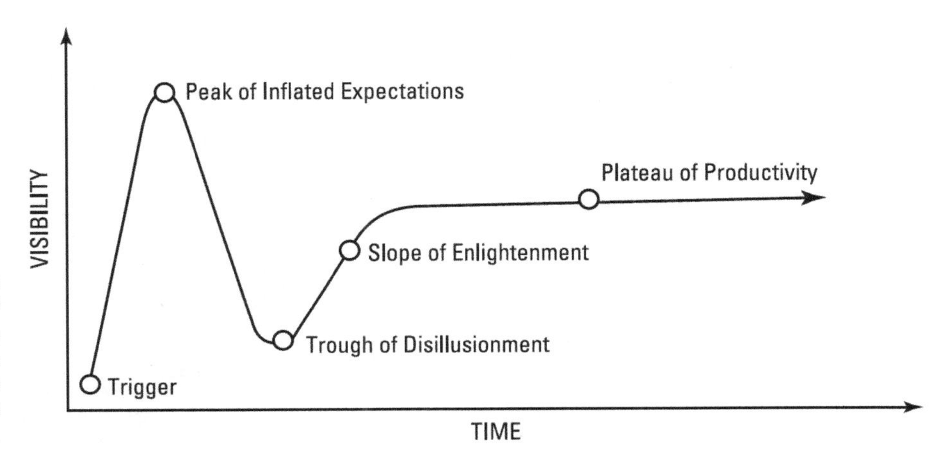

FIGURE 4-3: My version of Gartner's hype cycle, applied to the initial phases of DevOps adoption.

Expect a surge of excitement at the beginning of your mission. If you can get through the trough of unexcited (and often messy) conversations about what exactly is needed, you'll begin to see the progress you want. By then, you'll have the executive buy-in, engineering groundswell, and management adoption needed for your DevOps transformation process to enter a steady state.

Don't give up. Feeling frustrated is natural. You might even think about quitting, which is also normal. Going with the flow and allowing inertia to determine your future are so much easier. But making your job awesome is your job, and I believe in your ability to transform your organization in a meaningful way for you, your colleagues, and your customers.

Responding to pushback

Pushback will be a natural part of taking on the challenge of transforming your organization's culture into a DevOps organization. Sometimes the pushback is quiet; at other times, it's loud. No matter how the pushback happens, expect that people will push back against the idea of DevOps from all departments and groups of the company: sales, marketing, engineering, business stakeholders — you name it. The reasons will vary. Some people will have valid concerns; other reasons will be absolutely outlandish. Most will be rooted in fear.

Navigating the chasm

Earlier in this section, I present an adoption curve (refer to Figure 4-2). Geoffrey Moore popularized the adoption curve and highlighted the most vulnerable portion of the innovation adoption life cycle in *Crossing the Chasm*. The chasm is a portion of the adoption curve between the early adopters and the early majority. This is where you'll experience a tipping point of adoption just as you reach the Trough of Disillusionment in the Hype Cycle (refer to Figure 4-3). Dealing with this chasm, depicted in Figure 4-4, is perhaps the most challenging portion of DevOps adoption. At this point, you experience either full executive support or a groundswell of engineering excitement but have yet to hit majority adoption.

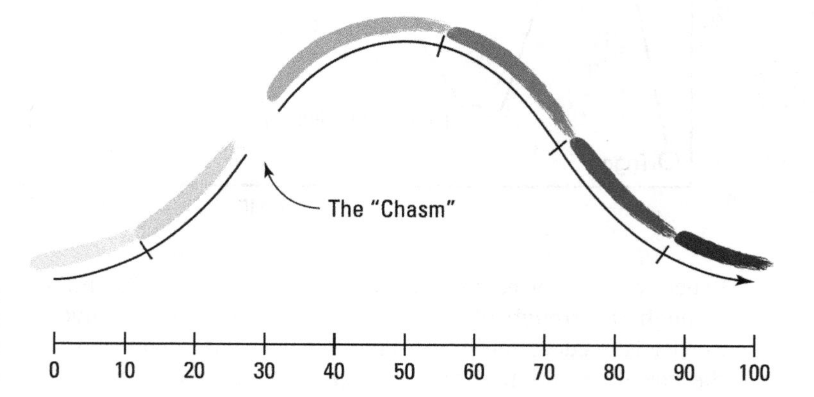

FIGURE 4-4: The chasm in the DevOps adoption curve.

Early adopters enjoy being first, and because of that added advantage, they don't care as much about the details. On the other hand, the folks in the early majority want to know that DevOps actually works. They may need some additional evidence. If you get stuck in this chasm, I recommend that you take your small band of innovators and early adopters and start practicing DevOps yourselves. Encourage engineers to experiment with the possibilities. Show them how much DevOps can improve productivity and collaboration. Most of all, don't get discouraged. Rely on the information this book arms you with. Part 2 looks at the entire software delivery life cycle linearly to equip you with what you need to inject DevOps along every stage of software development. In Part 3, you see how to connect the circuit and transform that linear pattern into a cycle of continuous improvement focused on the customer. Along the way, you discover everything you need to move from the persuasion phase to the implementation phase of a DevOps transformation.

Asking "Why?"

You may have heard about a technique known as the 5 Whys, an exercise to uncover the root cause of a problem. This exercise was — surprise! — developed at Toyota. It's often seen in kaizen, lean manufacturing, and Six Sigma — all of which are approaches to project management. Although the "root cause" concept is an antiquated approach to post-incident reviews (refer to Chapter 18), the 5 Whys technique is still a useful way of thinking through problems.

For an example of using the 5 Whys technique, imagine that someone expresses doubts over DevOps. Here are some "why" questions to ask, followed by possible answers (you're not restricted to just five):

- **Why are you hesitant to adopt DevOps?** Because I've seen Agile fail. What's the difference?
- **Why did Agile fail?** Because we went through the motions but never truly embraced an agile approach to software development.
- **Why did your team struggle to become agile?** Because sales and marketing determined the release schedule and we had no insight into customer feedback.
- **Why couldn't you talk to other departments or customers?** The product owners acted like gatekeepers and everyone stayed in their silos.
- **Why couldn't we learn from that failure and invite sales and marketing to our team meetings?** That might help them get insight to the challenges of engineering.
- **Why don't we use feature flags to ensure that products are released at a regular cadence for sales engineering but that we can adopt for continuous delivery?** That might actually work. (Refer to Chapter 11 for more information on continuous delivery and feature flags.)

Rarely do problems present themselves in an obvious way. Instead, someone may appear to be hesitant about DevOps but is actually worried about automating themselves out of a job. Or they don't want to put energy into something, only to have a manager tell them "no." Digging into the underlying fears that buttress your opposition gives you insight into how to best address the concerns and unify the team.

If you come up against someone who is vehemently against DevOps — either subverting your efforts or openly challenging you — don't take those reactions or challenges personally. They're most likely driven by fear: of the unknown, of failure, of success, of becoming irrelevant. Showing empathy for that person's fear and gently trying to discover the root of it can persuade all but the most cynical engineers on your team.

» Quantifying DevOps

» Creating DevOps case studies

Chapter 5 Measuring Your Organization

When considering how to make improvements in your organization, you can easily get a bit overwhelmed or, after deciding on a plan, you can want to jump in all at one time. It's a bit like setting a New Year's resolution to lose 15 pounds: You're tempted to cut calories dramatically and head to the gym every day. Although such an approach may seem ideal, it's likely to be unsustainable. For most people, it's too much change, too fast. DevOps transformations are a bit of the same. You have to leverage small wins and build momentum.

In this chapter, I suggest ways to get a baseline idea of where you're starting from and track your progress as you continue implementing DevOps. You also find out specific questions for employee surveys and understand the difference between quantitative and qualitative case studies.

Measuring Your Progress

A popular meme is floating around the Internet about the difference between what you think success looks like and what it looks like in practice. In Figure 5-1, you find my version of that meme. Your DevOps transformation will not be a straight line to success. You will have victories, setbacks, and headaches. At some moments, you'll want to throw your computer out the window. (But you're an engineer, so you're used to this.) Just keep with it.

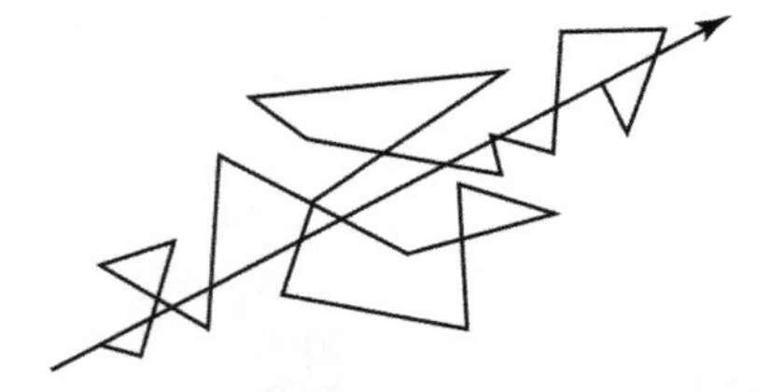

FIGURE 5-1: The picture of success.

Before you start adopting practices and implementing changes, you want to be sure to have a baseline from which to measure your success. This idea is similar to some types of medical tests. Every year when you go in for your annual checkup, you most likely get your blood drawn. Your doctor doesn't order this test because something is wrong, but rather to establish your baseline numbers for comparison year over year. That way, if something jumps or drops unexpectedly, you know what's "normal" for you and what needs additional follow-up.

I'm extremely hesitant to list a series of key performance indicators (KPIs) for you to track. The reason for my reluctance is Goodhart's law. Named after economist Charles Goodhart, this law states that when a measure becomes a target, it ceases to be a good measure. Goodhart wrote about the topic in his 1981 paper, "Problems of Monetary Management: The U.K. Experience." (He included this paper as a chapter in his book, *Monetary Theory and Practice,* in 1984.) He stated, "Any observed statistical regularity will tend to collapse once pressure is placed upon it for control purposes." (I like the layperson's version better.)

This idea contrasts with the thinking of Peter Drucker, a famous American management consultant, who consistently stated that what isn't measured can't be managed. The clash between Goodhart and Drucker leaves you and everyone else in a bit of a crunch. Should you measure or not?

I think there's a tension between these two positions that is right where you want to sit. It's a bit like tight-rope walking on a piece of floss. You're going to fall off occasionally. I have a friend, Reverend Jasper Peters, who always says we should hold things with an open hand. I really like that phrase. If you've ever played with fireworks, you know the difference between a closed palm and an open palm if something goes wrong and a firework goes off in your hand. One leaves you with burns; the other takes off your hand. The same applies if you've ever been on the receiving end of a punch. (This got dark.)

But the answer to whether you should measure your progress is yes. You should track performance metrics, too. But hold those measurements with an open hand. In other words, use them as points of reference that give you some perspectives (among others) of your success or failure.

I do list some KPIs in the next section that you can consider tracking when you first get started. As I say elsewhere, however, you should never use these KPIs to measure individual or team performance in a review of any kind. Nor should you tie bonuses or other monetary incentives to these measurements. That'll get you bad results in a heartbeat. Also, like the Constitution, these KPIs are akin to a breathing, living document. They should evolve.

Do not feel limited by suggestions, nor should you feel that you must track each and every one. They are a sampling upon which to build your internal DevOps culture and measure your team's progress. Add, remove, play, experiment. Have fun! (And if you just snickered at the thought of work being fun, we have work to do, my friend.)

Quantifying DevOps

If you're unsure of the meaning of any of the terms in this section, don't fret. Nor should you worry if you aren't sure whether your company needs to improve in a certain area, or how to implement a change. The information in this section is meant to be a starting list for you to begin to track your progress.

Note that I divide potential measurements by people, process, and technology. This is the tripod of DevOps, and you'll see this pattern repeat itself in this book and within the DevOps community.

People

Your team should be your first priority. Ensuring that they're happy and fulfilled with their work, as well as that they're using their time productively, should drive your initial data collection. But don't forget about your customers! They do pay the bills, after all. You want to ensure that customer satisfaction ranks high and stays high.

- **Employee satisfaction:** Survey the team. Are they happy? What do they love about their job? Where do they see room for improvement? Keep it anonymous and keep it open-ended. Allowing people to comment in a free-form manner in the beginning will help inform you on what you should be tracking in a more quantitative sense throughout the process.

- **Average meeting cost:** Engineers are expensive. Try running this experiment: Next time you have a requirements meeting or a sprint planning meeting, add up the estimated hourly salary of everyone in the room. Then multiply that number by the total time of the gathering. The number will be big. Endless meetings are a sign of poor collaboration, distrust, and an ineffective process. You can never eliminate meetings entirely, but watch the length of time you take engineers away from their desks. If such activities aren't adding value, cut them.
- **Customer usage:** How many users did you have sign up this week? How many cancelled their accounts? Do the cancellations track against any new feature release, or an outage? What features do customers use the most? Do you have some features that almost no one uses and that should be deprecated? Maintaining code is expensive. Here are a few terms regarding customer usage that are worth keeping an eye on:
  - **MRR:** Monthly recurring revenue
  - **MRR churn:** Monthly recurring revenue lost from customers who cancel
  - **Contraction:** Customers who downgrade their paid plan
  - **Expansion:** Customers who sign up for a more expensive plan
- **Number of customer tickets:** Typically, customers call only when something's wrong, so the number of calls is a good general measure of how intuitive your site is and how good your documentation is. Find out which areas of the site are difficult to use or which features are the least helpful. Identify which services are brittle or slow.
- **Customer satisfaction:** Sometimes referred to as CSAT, customer feedback is a key indicator for you. Determining your customer satisfaction can involve simply asking customers whether they're satisfied with the overall service or if they felt happy with the level of support received during a customer support call.

Process

Procedure drives much of your daily work. After you measure people, measuring the processes you've developed as organizational habits will help you determine where you're succeeding and where you need improvement.

- **Deployment frequency:** Do you deploy every day? Multiple times a day? Maybe every week or every month? Every . . . *(shudder)* year? Often, the continuous delivery approach is the lowest-hanging fruit at a company when they first decide to adopt DevOps as an engineering process. (I tell you about continuous integration/continuous delivery, or CI/CD, in Chapter 11.)

- **Size of deploys:** Tracking the size of your deploys is tied very closely to deployment frequency. Typically, infrequent deploy schedules hint at large deploys. The larger the deploy, the more likely it is for something to go wrong, and the harder it will be to identify what exactly may have caused the error. Small, frequent deploys are ideal.
- **Deployment length:** How long does actually releasing software to your customers take? Seconds? Minutes? Hours? Is it a manual process? Can developers release their code to production or does someone from operations have to initiate a deploy? I dig further into the topic of speeding up your deployment time in Chapter 11, but you want to automate deploys as much as possible. It eases the burden and removes some opportunity for error.
- **Defect escape rate:** How many bugs do you find in production after going through automated testing and a review by QA?
- **Recurring failures:** How often do bugs show up twice (or more)? Recurring failures are a sign of bugs slipping through the cracks. It could be that bugs aren't tracked well, aren't fully fixed, or aren't thoroughly tested.
- **Lead time:** How long does your team take to develop software? In other words, how much time passes between when you start work and when you deploy to production?
- **Mean time to detection (MTTD):** How quickly do you determine that something went wrong? Waiting for 100 customers to notify you on Twitter that your site is down is not an ideal way of discovering a problem. MTTD measures the time from when a problem begins to impact customers to when you discover it.
- **Mean time to recovery (MTTR):** Related to MTTD, MTTR averages how long you take to recover from a failure from the time it began to impact customers to the time you put a fix in place. MTTR uses an arithmetic mean, which assumes a normally distributed data set. The flaw of MTTR (and using any one measurement to evaluate performance) is that one major incident can make your MTTR plummet and skew your data inaccurately.

Technology

The technology and automation tools that you utilize in your system will determine the remainder of what data you should track, including test coverage, availability, reliability, error rates, and usage:

- **Automated test coverage:** How much of your application is tested? Are all the tests valuable in that they test something real? Does your test suite include only happy path tests (verifying expected functionality with expected inputs) or does it also include sad paths (validating how a function handles unexpected behavior)? Does someone write a test every time a bug is fixed?

- **Availability:** What is the uptime of your application? Often, companies have a service-level agreement (SLA) with customers that addresses uptime and availability. Are you meeting the expectations set by your SLA?
- **Failed deploys:** How many deploys go awry? How many cause incidents? Do they ever cause an outage? Which services are affected if a deploy causes service to be disrupted? Are you prepared to roll back any deploy quickly?
- **Error rates:** How many exceptions get thrown in production? It's a good idea to track database connections, time-outs, and other errors. An application performance management (APM) tool can help you identify which areas of your application are providing a suboptimal experience for your customers. Datadog, New Relic, Dynatrace, and AppDynamics (along with other competitors) all provide APM services.
- **Application usage and traffic:** Along with error rates, application performance management (APM) can help you track how much traffic your site is experiencing. Often, a surge of traffic or a sudden dearth is a sign that something might be wrong. As microservices (covered more in Chapter 20) become more popular, it's important to track dependencies. One critical service can impact others and have a cascading impact on your site's availability.

Collecting the data

It's extremely common for engineering teams to have absolutely zero data on their current performance. How long does a deploy take on average? No one knows. What's the monthly recurring revenue of your most popular application or service? Anyone's guess. What's the average weekly cost of meetings on your team? Uhhh . . .

If what I just described sounds a lot like your current team, don't worry too much. Again, you're in the majority of teams. But you don't want to be average, do you? You want to be the best. And to be the best, you've got to measure your actual output. You need to track your performance as an engineering team.

DevOps emphasizes metrics not as a measuring stick against some abstract version of success or failure but instead to inform you on how to keep making continuous improvement.

I recommend automating as much data collection as possible. You should also collect as much data as you can afford. You don't have to start collecting data on every metric described in the previous section tomorrow. Such a goal would be overwhelming and likely impossible. Instead, pick one to three metrics to focus on and work on setting up automated data collection.

These analytics will inform your continuous improvement. You can track them slowly over time and see how far you've come since you started.

If you still have some folks in your company who need convincing of the effectiveness of DevOps, this data will be an absolutely priceless tool for winning them over.

MAKING INCREMENTAL CHANGES

DevOps transformation is not an overnight process. After you begin applying DevOps principles, it will be weeks or months before you see measurable progress. Just as you wouldn't expect to lose 15 pounds overnight (that would be more concerning than elating), you shouldn't expect to see massive changes in your organization too quickly. But after you hit a stride, you're likely to see consistent improvement.

Many of the foundational principles of DevOps — trust, rapport, respect — take time to build. You can begin to influence this behavior through process, but much of it requires space and time — for your employees to step away from their desks, get to know each other, and talk about things. Some of the topics they talk about will relate to work; some of them won't, but all of it will be valuable.

Think about having to receive bad news. Say that the project you've been working on for three days needs to be scrapped. How would the conversation go if your best friend told you about this situation? What about a stranger? Chances are that the conversation with the former would be much more respectful than with the latter. When you have rapport with someone, it's easier to not take things so personally and instead focus on the facts. You don't feel the need to defend yourself because you know you're safe. Your friend knows you and loves and accepts you. That's the mindset you want your engineers to have with each other.

Of course, not everyone's going to love and adore each other. People are people, and some people just don't get along. You can, however, inspire mutual respect and understanding in any scenario.

Developing internal case studies

One approach that can be extraordinarily helpful in building up morale internally and showcasing improvements externally is to create internal case studies. If you decide to go this path, the impact will far outweigh the time you invested in building the case studies.

You can create a case study out of literally any metric. The general process involves choosing a metric that you want to measure, tracking progress, establishing your current baseline, and collecting data as you slowly improve your performance.

I highlight two potential case studies in this section. I mean these not as a prescription but rather to inspire you to think about how you can group certain metrics and begin to link the impact of one activity on another. As you begin to look at your engineering organization more holistically, you'll start to see just how much influence one activity has on others. Negative cascading effects can cost your team in morale, time, and resources, not to mention their impact on the customer.

A qualitative case study: Focus on your employees

For a qualitative case study, you focus wholly on your engineers' satisfaction with their jobs and perceived level of collaboration.

Measure employee satisfaction

To measure employee satisfaction, create an open-ended survey and send it to your employees. The first time you do a survey like this, give ample opportunity for employees to speak their minds freely through comments to help you uncover areas that are ripe for improvement but that you may not have expected.

Following are questions to start with, but tailor them as you see fit. Be sure to emphasize that the survey is anonymous. Ideally, no one will be able to tie specific answers to an employee. If avoiding that situation is absolutely impossible, opt to have a single person oversee the process of removing identifying information.

Here are the questions I suggest you ask:

- On a scale of 1–10, how do you rank your pride in working at this company? What would make you feel more proud, inspired, or happy at work?
- On a scale of 1–10, how do you rank your feelings of empowerment and autonomy to make decisions at work? What would improve your score?
- On a scale of 1–10, how do you rank your supervisor's performance? What would improve their ranking?

- On a scale of 1–10, how comfortable do you feel asking for help when you need it? What stops you from asking for help? What would make you feel more comfortable?
- On a scale of 1–10, how do you rank this organization's leaders at informing you about mission, vision, and values? How could this be improved?
- On a scale of 1–10, how satisfied are you that you receive appropriate recognition for good work? What else do you want us to know?
- What do you think is working well?
- In which areas do you see room for improvement?
- Is there anything else you think this survey should have asked?

Calculate average meeting cost

Over the course of two to three weeks, track the time spent in various meetings. Estimate the average hourly salary of everyone in the room and multiply that number by the number of hours you spend gabbing to each other in a conference room. You're only estimating, and you don't need to have actual salary information. The purpose is to discover a baseline of meeting costs, and estimations will serve you plenty well enough.

Also, your goal here isn't to eliminate meetings. Some level of communication is critical to passing information effectively. You're extremely likely, though, to have plenty of meetings that don't create impact, either for your engineers or your customers. Productive meetings should create positive output — for example, clarified requirements and key architecture decisions.

Track development lead time

The goal of tracking development lead time is to establish your current baseline and then slowly reduce it. This type of tracking may apply more to the long term, but you can likely get a pretty good idea of development lead time by looking at single features created by the team.

Look for bottlenecks along the process so that you can more easily identify how lead time can be reduced. Here are the questions to consider.

- Are overarching architecture decisions understood by everyone on the team?
- Are requirements clearly stated and is context communicated to the individual developers?
- Do junior engineers need more training on specific tools?
- Would code reviews or pair programming increase velocity?

- What's the process that a feature must go through after it has been developed? Is there testing? A security review?
- How often do features get kicked back to the developer from testing or security?
- Is the code sufficiently well documented that a second engineer could pick it up if necessary or does the original engineer have to be the one to finish?
- Is code deemed production-ready deployed immediately? Or is it held in a queue for a larger release?
- Can developers deploy their own code or do they rely on an operations engineer?

A quantitative case study: Home in on deployments

This case study is much more quantitative than the last. It looks at raw numbers to give you a better idea of your team's performance in relation to deploying software to production. Specifically, how often do you deploy? How long does a deployment take on average? What's the average size of the release?

Or, collect data on deployments going forward. If you're using any kind of release software, such as Jenkins, you likely have (at least) weeks of data on past deployments. If you don't, set up some type of tooling to help you automatically collect deployment analytics. Here are questions to consider:

- What is the average deployment frequency? Days, weeks, or months?
- What is the average size of a deploy? How many features or services are impacted? Does the deploy typically affect only a single portion of the codebase or does it often include large, sweeping changes?
- Can you tell whether a release goes poorly? Does alerting or some way of confirming the new software was successfully deployed take place? How can you tell whether your application looks and behaves the same way? Do you feel confident in your testing process or do you often have people click around the site after a deployment to make sure it looks all right?
- What time of day is software released? Do you have a set time? Do you require engineers to deploy during off hours? If so, how often are you asking people to work in the evenings or weekends for planned deploys?
- How often do you have to roll back a deploy? Or create a hotfix? Is your team prepared to manage problematic releases? If you can roll back, how long does it take for the revision to take effect?

2

Establishing a Pipeline

IN THIS PART . . .

Think about the software development life cycle as a linear process throughout which you may optimize with DevOps by addressing concerns earlier in the process and beginning a CI/CD practice.

Invite everyone to the planning table when first gathering requirements and designing features for a new product or service.

Architect your system to be flexible and resilient, and document design decisions as you work.

Choose specific languages, frameworks, and programming patterns to develop well-written code that is more easily understood and maintained.

Automate testing to utilize every type of test and ensure that code is functional across multiple environments.

Take CI/CD to the next level and release software using deployment strategies proven to facilitate small, frequent releases of code with increased service availability.

» **Shifting operations to earlier ("left") in the development cycle**

Chapter 6
Embracing the New Development Life Cycle

In this chapter, I describe what's often called the software development life cycle, or pipeline. Although some nuanced differences may exist between the two concepts (depending on whom you ask), I use development life cycle and development pipeline interchangeably.

The tech industry uses the term *software development life cycle* (SDLC) to describe the process from creating an idea for a new product, application or feature to actually deploying the new software to customers in a production environment. I actually prefer *delivery* over *development* because that word removes any implication that developers are the star player in the software life cycle, which would reinforce the old ideas of silos and divisions between developers and operations people.

Many iterations of the development life cycle exist, with various steps, and some involve more steps than mine whereas others involve fewer. In this chapter, I explain how DevOps changes the approach of the development life cycle. I also briefly explain the various phases of that life cycle, each of which is covered in separate chapters throughout this part of the book.

Inviting Everyone to the Table

The most important purpose of creating the development process is that it provides a framework for everyone to work within. Your engineers won't necessarily fit perfectly within one stage of the pipeline and only do that one bit of work; that scenario would just be creating more silos, with the engineers in one section simply doing their work and lobbing it over to the next section. That's the exactly opposite of what you're trying to build.

Instead, you create a recipe for success for your team: a way of breaking down the development process like an algorithm — or recipe — so that everyone understands how your company and your DevOps culture develops the best software and delivers it to your customers quickly and reliably.

This pipeline framework that you'll develop is a process through which all your engineers can learn new skills and pitch in at various stages. The most important benefit of the development pipeline is that it invites everyone to the table. It gives everyone the opportunity to get involved as they see fit and to learn new skills if they're interested. It also gets your team using a common language. You'll be able to discuss the same concepts using the same words, which is vital for smooth communication.

Figure 6-1 shows a software development life cycle drawing often seen in DevOps.

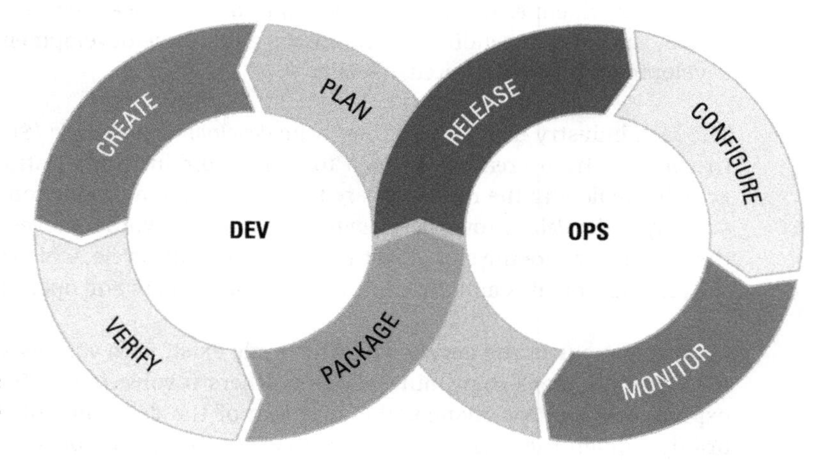

FIGURE 6-1: The DevOps tool chain.

Changing Processes: From a Line to a Circuit

Development processes have changed radically over the last few decades, and for good reason. In the 1960s, Margaret Hamilton led the engineering team that developed the software for the *Apollo 11* mission. You don't iteratively launch humans into space — at least they didn't in the 1960s. It's not an area of software in which "fail fast" feels like a particularly good approach. Lives are on the line, not to mention millions of dollars.

Hamilton and her peers had to develop software using the waterfall methodology. Figure 6-2 shows an example of what I think of as a waterfall development process (occurring in a straight line), and Figure 6-3 adds the phases. Notice how the arrows go in one direction. They show a clear beginning and a clear end. When you're done, you're done. Right?

FIGURE 6-2: Drawing the line of waterfall development.

Nope. As much as many people would like to walk away from parts of their codebases forever (or kill them with fire), they usually don't get the privilege.

The software developed by Hamilton and her team was a wild success (it still blows my mind that they developed in Assembly with zero helpers like error messaging). Not all projects were equally successful, however. Later, where waterfall failed, Agile succeeded. (As mentioned in Chapter 1, DevOps was born out of the Agile movement.) Agile seeks to take the straight line of waterfall and bend it into a circle, creating a never-ending circuit through which your engineering team can iteratively and continuously improve. Figure 6-4 depicts how to think of the circular development life cycle.

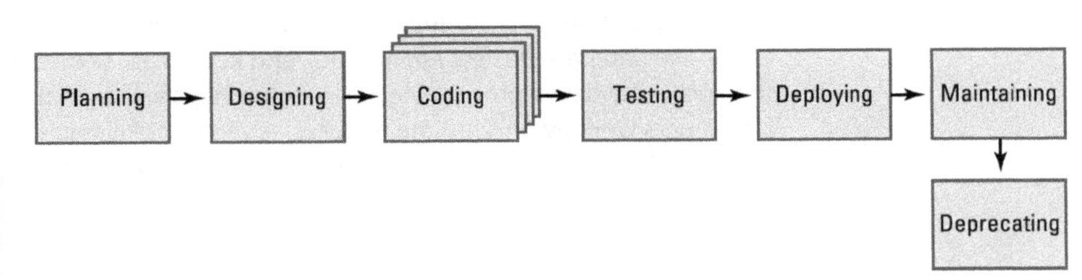

FIGURE 6-3: The waterfall development pipeline.

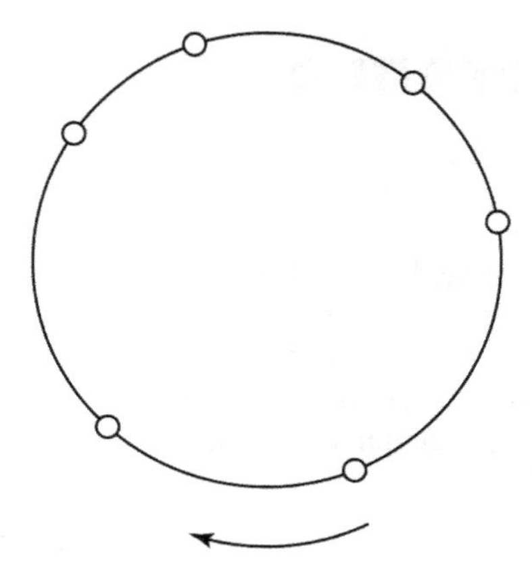

FIGURE 6-4: Creating a circuit.

Often, the various loops prescribed by different organizations are influenced by the products those vendors sell. For instance, if the vendor sells infrastructure software and tooling, they likely emphasize that portion of the development life cycle, perhaps focusing most on deploying, monitoring, and supporting your software.

I have nothing to sell you. The stages I focus on are the ones that I saw as being the most critical as a developer, along with the ones I see people struggling with the most as I teach organizations to better manage their software development and adopt DevOps.

The six stages of the software development life cycle that I highlight in this part of the book (Part 2) are

- **Planning:** The planning phase of your DevOps development process is perhaps the most key to your DevOps mission. It sets you up for success or failure down the road. It's also the most fertile time to bring everyone together. By everyone, I mean business stakeholders, sales and marketing, engineering, product, and others. Chapter 7 covers the planning phase.
- **Designing:** In most companies, the designing phase is merged into the coding phase. This monstrous amalgam of design and code doesn't permit a separation of the architectural strategy from implementation. However, if you leave things like database design, API logistics, and key infrastructure choices to the end of the development pipeline — or, perhaps worse, to the individual developers working on separate features — you'll quickly find your codebase to be as siloed as your engineering team. Chapter 8 covers the designing phase.

- **Coding:** The actual development of features is the face of the process and gets all the glory. But I argue that it's one of the least important steps in your development life cycle. In many ways, it's simply the execution of the preceding areas of your pipeline. If done well, coding should be a relatively simple and straightforward process.

  Now if you're a developer and just gasped at that last sentence because you've dealt with hundreds of random and difficult-to-solve bugs, I know how you feel. Coding is hard. Nothing about software development is easy. But by mastering the planning, design, and architecture (and separating them from the actual implementation of code), you ensure that the hardest decisions of software development are abstracted away. Chapter 9 discusses the coding phase.
- **Testing:** Testing is an area of your pipeline in which engineers from all areas of expertise can dive in and get involved, enabling a unique opportunity for learning about testing, maintainability, and security. There are many different types of tests to ensure that your software works as expected. Chapter 10 covers various types of tests for this phase.
- **Deploying:** Deploying is the stage that is perhaps the most closely associated with operations. Traditionally, your operations team would take the code developed by your developers and tested by your quality assurance (QA) team and then release it to customers — making them alone responsible for the release process. DevOps has had an enormous impact at this phase of the development process. Also, deploying is one of the areas from which to find the most automation tools to pull. From a DevOps perspective, your priority is simplifying the deployment process so that every engineer on your team is capable of deploying their code. This is not to say that operations doesn't have unique knowledge, or that operations teams may be disbanded.

  Operations folks will always have unique knowledge about infrastructure, load balancing, and the like. In fact, removing the manual task of deploying software from your operations team will allow them to save you time and money elsewhere. They will have the time to work on improving your application's reliability and maintainability. In Chapter 11, I discuss how to smooth out your deployment process and create a continuous integration and continuous delivery (CI/CD) environment.

The most important aspect of a delivery life cycle within the DevOps framework is that it is a true loop. When you get to the end, you go right back to the beginning. Also, if you receive support feedback from customers at any point along the way, go back to a subsequent phase (or the planning stage) so that you can develop software in a way that best serves your customers.

Think of the content of this part of the book as two-dimensional, and the content of Part 3 as three-dimensional — the evolution of that pipeline or delivery life cycle.

The first part of building a pipeline is to treat it linearly. You are building a straight line with set stages and checkpoints along the way. Within this framework, you can view the software development life cycle as something you start and something you finish. Waterfall lovers would be proud.

But reality doesn't let you work in a straight line. You can't just start producing code, finish, and walk away. Instead, you're forced to build upon the foundational software you released on your first iterative loop and improve it through the second cycle. And so on and so on. The process never ends, and you'll never stop improving.

This book helps you connect the start and finish of that straight pipeline so that you begin to understand it as an entire circuit, or loop, for you to continuously develop and improve.

DEPRIORITIZING TECHNOLOGY (IT'S NOT HERESY!)

At every stage of this development life cycle, you will find a dozen or so odd tools all claiming to be absolutely necessary to your success at that particular stage. Don't get me wrong: Tools are incredibly useful. This is why I've dedicated an entire section to tooling your pipeline (which you can find in the chapters in Part 5).

Unsurprisingly, the tools are the least important aspect of building your development pipeline. The most important facet of this process is that it is continuous. At no point along this life cycle is your team stagnant. The fact that one engineer is releasing software to customers doesn't mean that all your developers have stopped coding. Instead, everyone continuously develops, tests, deploys, and improves your software. DevOps focuses on continuously improving and creating a pipeline, with an emphasis on continual flow.

For now, don't stress about the tech you'll use. Instead, stay focused on your people and the processes you're building to better support those people in their work.

Shifting Ops "Left": Thinking about Infrastructure

"Shifting left" as a term first appeared in the 1990s when people realized that waterfall development created inferior software for the market, and products that often required expensive fixes. The problem was that testing was too far to the right, or late, in the software development life cycle. This realization doesn't just apply to testing anymore. It's important to shift ops (and other specializations) left, too.

If you dig into other DevOps literature, you occasionally come across the phrase "moving left" in regard to teams like operations, security, and quality assurance (QA). This idea simply refers to moving the work completed by these teams leftward in the development pipeline, or sooner in the process. Traditionally, the work of operations was the last thing anyone thought of. Most of the organizations I have worked for have involved operations only after code has been developed. This situation is unfortunate because it strips operations engineers of their ability to properly plan and design infrastructure to support the code.

Many failures seen in production are expensive, typically costing $5,000 per minute. The cost of your production outage will vary, but it's expensive no matter how you cut it. Often the cause of an outage is a lack of consistency in your infrastructure as well as the development process. When you bring operations into the conversation early, you give them the opportunity to use their area of expertise to inform the rest of the team on things to look out for and how to best prepare for the successful deployment of software.

Shifting ops left refers largely to a philosophy of prevention rather than reaction. You don't wait to detect a failure and then try to fix it. Instead, you think through the potential failures of the system and do what you can within the constraints of your resources to prevent unfortunate surprises at the end of your delivery life cycle — when those potential failures are most likely to impact customers.

Automated continuous testing is a critically important aspect of this approach. Everyone on your team, especially developers creating new code, should be running your automated test suite throughout the entire development process. I cover how to create an automated test suite in Chapter 10, but for now, remember that taking the time to write tests will save you hours of expensive headaches down the road.

Shifting deployments left, too

Deploying continuously — meaning that developers release their software as it's developed — is ideal for many teams. But continuous deployment takes a great amount of work to implement and do successfully. Don't take continuous deployment lightly, and realize that it's not right for everyone. I like to tell people to keep the idea of continuous deployment as a type of long-term goal. Like nirvana, the point isn't necessarily to actually get there or achieve it, but instead to work toward it and accomplish things along the way.

One way to involve operations earlier in the process of development is to have the operations team develop patterns and checklists to help developers design software ready for deployment. Often, operations folks have to go through a series of manual steps to deploy code into production. If you're not ready for automated releases, you should aim to transfer the steps into a checklist so that developers can validate that their code is ready for the production environment.

In addition to using checklists, you should build the patterns set by your operations team into your automated test suite. That way, developers don't have to necessarily "code better" but they can validate their work as they build it.

Automation eases the burden of shifting operations left in your software development life cycle. Automating the consistency of the deployment process will improve your confidence in each deployment. (Who likes to stress-sweat?) Make each deployment environment as similar as possible within the constraints of your resources. Do the same for development environments, testing environments, staging environments, and production environments — including cloud environments, whether public or private.

Mimicking production through staging

Almost all production environments are more robust than developing or staging environments. A development environment is what each of your engineers uses to run code on their machine as they build it. Development environments are typically the most lightweight of all the environments. The staging environment is what used to test (occasionally there is a testing environment as well) and validate software before it's released into production. Staging environments should have as much parity with production as possible.

» Gathering requirements

» Creating a minimum viable product (MVP)

» Designing personas to identify your customers

Chapter 7
Planning Ahead

DevOps was born of the Agile movement. In fact before Andrew Clay Shafer and Patrick Dubois decided on the term *DevOps*, Shafer preferred "agile project management" — a bit of a mouthful. (Shh! Don't tell Andrew I said that.)

Because you're reading this book, you're likely somewhat familiar with the Agile style of product management. You can think of DevOps as an evolution of Agile. It is an iterative process that allows you to plan, develop, and release code quickly. You adapt to changes faster in an evolving market, out-innovate competitors, and respond to failures at a more rapid pace.

This chapter aims to help you approach planning in a DevOps organization. Because of DevOps's origins in Agile, you'll notice many similarities to Agile if your company already works within that framework. If it doesn't, you might also encounter more friction in implementing DevOps — even in a highly tailored fashion for your company.

In this chapter, I discuss collecting product requirements, uniting the team around a shared vision, and understanding your constraints.

Moving beyond the Agile Model

Agile is so generally accepted today that it's hard to imagine a project management style before it. Whereas Agile's style is iterative, its predecessor's, waterfall, was linear. The waterfall model, described in Chapter 6, is a sequential series of events. In the days of shippable software — think ISPs on CDs — companies had to plan and develop software sometimes two years (or more!) ahead of a planned release.

THE ORIGINS OF AGILE

Agile was born from frustration with the waterfall model's inflexibility and constraining framework. In the 1990s, the software ecosystem changed and the management approach needed to change with it, resulting in the emergence of *scrum*, extreme programming and feature-driven development (some implementations of Agile principles). Although some of these methodologies originated prior to the signed Agile Manifesto, many people think of them as an offshoot of agile software development.

In 2001, 17 software engineers met and published the "Manifesto for Agile Software Development" (`https://agilemanifesto.org/authors.html`), which spelled out the 12 principles of Agile project management, paraphrased as follows:

- Satisfy the customer by continuously delivering beneficial software.
- Accept and welcome changing requirements along the process.
- Deliver working software frequently.
- Enable developers and business stakeholders to work in daily cooperation.
- Trust your engineers to get the work done.
- Convey information face to face.
- Realize that working software is the most important measure of success and progress.
- Maintain a constant pace of work.
- Strive for technical excellence.
- Simplify requirements and features.
- Allow teams to self-organize for the best product.
- Regularly reflect, as a team, on how to become more effective and adjust behavior accordingly.

The ultimate downfall of this approach was that it simply took too long. By the time the product shipped, the market had changed. Companies wasted endless amounts of money, resources, and time chasing an idea without any feedback from customers. In contrast, Agile prioritizes the customer and introduces the idea of continuous improvement. (See the earlier sidebar "The origins of Agile" for an overview of Agile principles.)

In DevOps, as in Agile, teams adapt to the ever-changing needs of the business or the market, and sometimes even the changing technology and tools. Teams set milestones, plan features, and develop code continuously.

In addition to its important role in shortening the long development cycle, Agile spurred companies into adopting a rhythm of continuously building products based on feedback from current customers. This rhythm consists of short periods of time called *sprints* that last usually no longer than two weeks, during which a team decides what work to finish and ship to customers.

Over time, though, the dramatic impact of Agile has lessened. In many ways, development teams had the greatest acceleration of productivity with Agile, whereas operations teams didn't see the same results. The lack of collaboration between these two sides of an engineering team only made the gap more evident.

In the wake of these frustrations, DevOps seeks to solve the problems left unaddressed by Agile. Above all else, it emphasizes the collaboration of all specialists in engineering — from product managers to testers to operations engineers.

Forecasting Challenges

Iterative or not, your product development has to start somewhere. The planning stage of any new product provides a unique opportunity to invite everyone to the table to share ideas and brainstorm —without the stress of approaching deadlines that occurs later on in a project.

REMEMBER

At the beginning of planning a new product, everyone is generally relaxed and open to suggestions. If your current culture doesn't reflect the openness you expect to see in the planning stage, I highly suggest that you apply a DevOps mentality to the problem. Ease the tension between people throughout the product development life cycle. Assumption of malice, lack of curiosity, and defensive egos are some of the greatest threats to your organization during your DevOps transformation.

If your company doesn't currently work within the framework of Agile, don't fret! It's never too late to adopt a new project management style. Don't let anyone persuade you that you're too late to the party to reap the benefits. You can absolutely continuously improve. Throughout the planning process, educate your colleagues on Agile and, specifically, the concept of sprints. Emphasize the importance of producing a lean minimum viable product (MVP) and breaking the work into smaller pieces. (See "Designing an MVP," later in this chapter, for details on how to design an MVP.)

Identifying project challenges and constraints

Every company has limitations — constraints around resources, compliance requirements, and market trends. Write down the constraints you must work within to get the project done. Ask your engineers to do the same, and then compare your answers. This exercise illuminates the different perspectives, motivations, concerns, and predictions of your entire team.

Every project is controlled by four constraints: scope, deadline, quality, and budget. You often hear an impossible-to-attribute saying that goes something like, "It can be done well, fast, or cheap. Pick two." Adopting DevOps requires you to integrate a project management style that agrees with the basic tenets of a DevOps philosophy. You should also tailor the general prescriptions in this chapter to your team and your specific constraints.

The two most common challenges facing any software project are schedule and budget. If you're a startup, leadership will likely want to be first to market with your project in order to gain the most market share. Venture capitalists (private investors who often fund tech startups) prefer companies that are fast and aggressive in their release cycle. If you're an enterprise, the constraints can become a little more interesting. You have other products and services that can't be impacted for the development of a new project. You may also have service-level agreements (SLAs), which dictate what your customers can legally expect to receive from your service, as well as compliance concerns.

WARNING

Every project has constraints, and you should see a huge red flag if a stakeholder can't list the challenges or limitations facing a company while undertaking a software project.

Schedule

What schedule limitations does your team face? The answer is never "none." Root out schedule constraints by interviewing people outside the engineering and product teams. For example, marketing likely has an event coming up in a few months at which they would like to demo something. Also, someone in sales may well have already talked to a customer who would be more than excited about your new feature or product. That customer might even want to be a beta user (someone who "test drives" software before a public release) to more rapidly produce user feedback.

You might also encounter scheduling constraints because of a financial goal or challenge. If you're a startup, leadership may want to raise another round in 6–12 months, which means that the product would need to be usable and in the hands of customers by then. Or maybe you just raised money and have a year's worth of financial runway (the time during which your company can sustain itself with the money in the bank). If your company is publicly traded, perhaps the business stakeholders want to schedule a product launch around an upcoming SEC filing.

REMEMBER

This part of the development life cycle is an information-gathering phase. You don't necessarily need to take action just yet on these constraints. Instead, you are creating a context at the start of your project that will inform your decisions as you move forward. If your schedule is tight, you can negotiate to have fewer features appear in the initial customer release.

Budget

Budgets are the second most common constraint. And for good reason, right? Most companies aren't Google or Amazon and made of money. Instead, they have to get creative and develop working software efficiently. (Not that Amazon isn't efficient, Mr. Bezos.)

The trickiest dynamic around budgets is that the budget constraints aren't always obvious. Often, businesses obfuscate actual budgets and costs to keep other parts of the organization from knowing about them. Or, in many large organizations, departments must fight for budget share once a year and then work within that budget until the next cycle. Here are a few aspects of budgeting to consider:

- **Head count:** Do you have enough employees to get the job done? If not, do you need to increase the number of engineers on a team? What is most cost effective? For example, are you better off with one senior engineer or two more inexperienced developers?
- **Infrastructure costs:** The new feature or product will likely need to be hosted. Keep in mind the costs of whatever hosting solution you choose, whether it's private, managed, hybrid, or cloud. I discuss moving to the cloud in Chapter 21.
- **Intersection of cost and time:** How much does it cost every week that you go over schedule? The number you calculate won't be perfect but it will help you make more educated decisions as you move forward in the project and come up against unexpected delays.

Gathering Requirements

DevOps emphasizes continuous planning, which is an Agile approach to integrating customer feedback into the future planning process throughout the project. But even in the most Agile scenario, a project almost always has a set of basic requirements for you to meet to fulfill the needs of your customers. I suggest you approach the requirements-gathering phase using three steps:

- **Share business objectives.** Create a product requirements document (PRD) that emphasizes the business objects of this project. This document should be one page and easily consumed by anyone in your company. The reason for a PRD is to provide the "why" and highlight the purpose behind the project. The shared understanding and infectious passion can carry the team during times of stress.
- **Create user stories.** Interview customers, if you haven't already. Include team members from design and engineering. First-hand experience with talking to customers can help bridge gaps of information that form when only product owners are allowed that contact. When you encourage a variety of people to ask questions and interact with customers, you build a much deeper understanding of the users' challenges and desires. I discuss gathering customer feedback in Chapter 13. The same approaches for contacting and interviewing customers after a product is released apply to asking for their feedback before you design a new service.
- **Set scope.** Ending up with an enormous pile of feature ideas at this stage is perfectly normal. In fact, it's great! You want people's imaginations working overtime and getting them excited. Allow everything to be added to the list. Then, refine it. Designing a lean MVP, discussed in the next section, will help you limit scope.

WARNING

After you've set scope, stick with it. You need a product owner who is capable of saying "no" to be in charge of evaluating potential features and deciding which get put into the product, and when. Although being agile in your development and continuously integrating feedback are important, make the process of adding new features rather difficult. Doing so discourages scope creep and prevents a thousand "great ideas" from being added at the last minute. You know what the project needs to accomplish, so do that and no more. Later, after it's released, you can always go back and add new features with each sprint.

Designing an MVP

A minimum viable product (MVP) is the bare minimum of a product that still accomplishes its most basic objectives without the excessive bells and whistles of additional features.

MVPs are critical to DevOps organizations because they don't require the significant up-front planning that a large-scale enterprise product did in the past. With an MVP, you don't work on a product for two years and hope that it succeeds when finally launched to customers. Instead, businesses can quickly test ideas and adapt to changing markets. If the first MVP doesn't quite hit the mark or wasn't received well, it didn't cost you much and you're able to pivot to another direction using the customer feedback you gleaned from the first MVP.

SUCCESSFUL MVPs

Many of the modern hegemonic tech companies started as extremely simple sites that did one thing. They just happened to do that one thing extremely well. By focusing on their most valuable aspect of the business, they attracted loyal early adopters who evangelized on behalf of the company. Two of those companies are Facebook and Airbnb.

Facebook

Mark Zuckerberg's product didn't start out as the verbose ecosystem it is now. It was a simple service that connected students by which college they had in common. Users could post messages to boards, and that was about it. Facebook didn't have chat, or photo storage, or timelines at its start. Instead, it launched as the most basic iteration of the product and attracted enough users to gain traction. After it had the users, the company expanded and added additional features.

Airbnb

The extremely popular rental booking site was born out of the difficulty its founders had in paying their expensive San Francisco rent. To try to make ends meet, the founders provided accommodation to friends who came into town, took pictures of their place, and advertised it. The idea took off, partly because one of the founders, Paul Chesky, lived exclusively through booked Airbnb rentals for a year.

If you're in the process of transitioning to a DevOps methodology, you need to get in the habit of developing MVPs for your new products. Removing excessive features and preventing scope creep are two of the biggest contributing factors of success in high-performing companies. Scope creep will absolutely murder team morale as more and more features are added at the last minute.

An astute planning process and hard lines around what can be added to the initial release will benefit the efficacy and longevity of your team.

Discovering the problem for your MVP to solve

Problems exist in almost every process. In fact, people often just ignore them until someone comes up with a solution. Only then, with the benefit of 20/20 hindsight, do they see the value of that solution. (One of the arguments against Ford's automobiles was essentially, "But what's wrong with a horse?")

You've likely already identified the problem in your process, or what you think is the problem. But the root issue often eludes people. If you've ever done a client consultation, you know that sometimes the problem isn't what people think it is. Often it's a symptom of a greater issue that lies beneath the surface. Do your best to unearth that issue. The closer you get to the root of the problem, the more successful your MVP will be. Here are some questions to consider:

- What is the challenge you want to solve?
- Why does the challenge exist?
- In which industry is it most commonly experienced?
- Does the problem affect the majority of people or is it niche?
- If you're the customer, why do you need this product?
- What's the value of solving this problem?

Identifying your customer

Your customers will drive every decision you make as a DevOps organization, but how do you figure out who struggles with the issue that your product is supposed to solve? In the "Determining Your Customer by Designing a Persona" section, later in this chapter, I talk about the importance of customer personas and how they can help you identify your customer to better understand how they think.

When you not only identify your potential early adopters — the people most willing to try new products — but also dive into their psyche, you help all the details emerge that will help you think and feel as they do.

If the problem affects you, awesome! That's extremely valuable insight to your customer — because your customer is you! (The problems you experience might be related to software, or . . . your dog. Every engineer is also a customer of a thousand other products.) Regardless, you need to speak to your customers. Even better is to hire engineers who have experience with your customer base or fit the profile of your customer themselves. Those engineers can then give you unique understanding as you design your product.

Scrutinizing the competition

If your company has no competitors, take that as a huge red flag. This idea may seem counterintuitive at first. Daymond John, founder of the company FUBU and investor on the TV show *Shark Tank*, says, "Pioneers are slaughtered and settlers prosper." It is almost impossible and incredibly risky to be the first company in a new industry space. It's much safer to enter an established market and differentiate yourself, just as Airbnb did. Airbnb may seem like a novel idea, but the problem it addresses, that of needing a place to stay, was already solved by hotels. Airbnb differentiated itself by giving homeowners the opportunity to earn extra cash and customers the chance to discover unique spaces in which to spend the night.

When you discover your competitors, dive deep into their product and their messaging. Here are some questions to consider:

- Who are their customers?
- Is your product in line with theirs?
- How will you differentiate your product to customers?
- Can you see any opportunities to steal customers? Or reach customers that your competitors have been unable convince?

Insight into the products already on the market can help you design the product with that context in mind and potentially avoid pitfalls previously experienced by your competitors.

Prioritizing features

It's great to dream big. In fact, I think it's really important. Dreaming big enables your entire team to really stretch their ideas and brainstorm everything that a product could be. This way, you invigorate your project and help ignite your team. When they're excited about the possibilities, they come up with unique innovations and points of view that can drive your product further.

A great exercise in product development is to list every feature you want to include in your product. Have everyone at your company do the same, merge the duplicates into single line items, and then stack-rank them — put them in order of importance.

In the end, you want to prioritize your list in order of importance. Consider which features you can't live without. Cut out any feature that isn't absolutely necessary to solve the problem you're attempting to mitigate.

If you're struggling to prioritize the potential list of features, make three lists:

- **Features you can't live without:** These are the items most vital to the product's capability to solve a problem for the user.
- **Features that are nice to have:** These are the ideas that will improve the product but aren't critical to its overall efficacy.
- **Features that don't matter:** These are the items that no one on your team is willing to fight for. They're ideas that simply died on the vine. It's nice to have a list to pull from in the future (who knows when your iterations will demand such a list of features?) but for now, tuck this list into a desk drawer and move on.

Designing the user experience

In this section, I'm not referring to building high-fidelity wireframes just yet. That's a bit down the road(map). Instead, in this part of the planning process, you should think about how the user will interact with your product. What is the main driving force that will bring users to the site and keep them there? What's the main activity will they be completing? What are the steps or tasks involved in that process? How will the user flow from one activity to the next?

Imagine that your product is a photo storage MVP. You want to think through the user flow and which features will enable the user to step through each phase of using the product. Take your list and categorize the features by the steps your user will go through in the flow of your product. Figure 7-1 demonstrates this process.

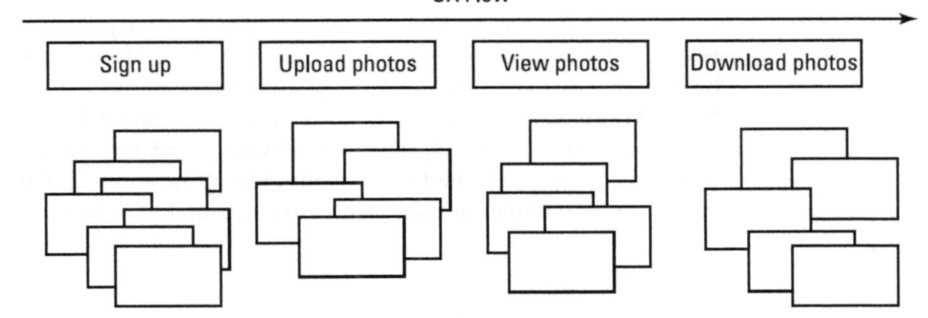

FIGURE 7-1: Organizing features by user experience (UX) flow.

After you have the features sorted by category, you need to order the features within each category by importance. Remember, a feature that you love might not be the most important. Emotional attachment to the ideas you have is much less important than the pertinence that a particular feature has to your users. How does it help them? Is it vital to making the MVP a usable product for customers? Order the feature ideas within each category, as visualized in Figure 7-2.

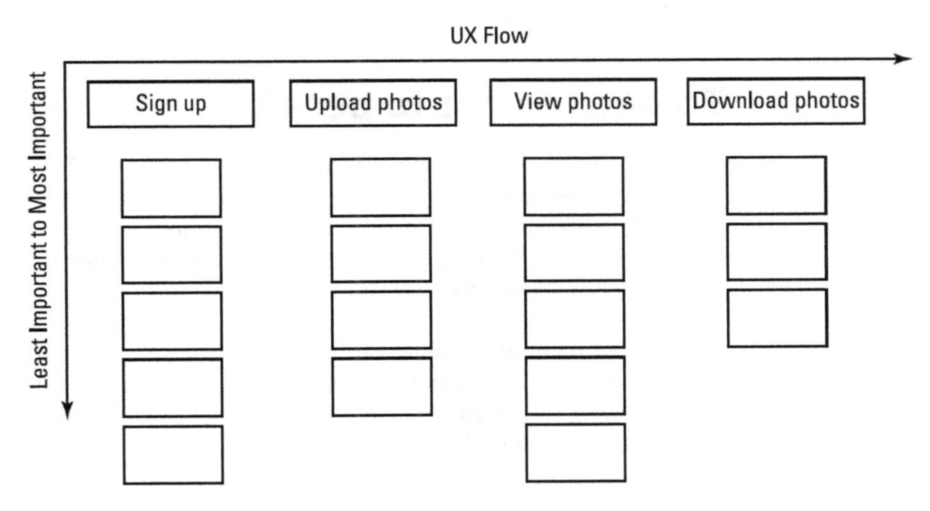

FIGURE 7-2: Prioritizing features by user experience (UX) flow.

Testing your hypothesis

When you feel comfortable that you've planned well, talked to customers, and integrated their feedback into your product plan, it's time to get building! Remember, watch out for scope creep (see the "Gathering Requirements section, earlier in this chapter). Only the must-have features — the ones that your MVP won't function without — should be integrated into this first iteration. The point isn't to necessarily knock the socks off your users. Instead, it's to prove the viability of your product.

In fact, in a few years, you should probably be pretty embarrassed about your MVP. That's ideal because it means that your MVP was stripped down enough to be a legitimate MVP. When you adopt DevOps, you accept that your development life cycle will be an iterative process. You will build something, release it, listen to feedback from customers, and then iterate based on the feedback. This cycle continues throughout the life of the product. It never ends. Figure 7-3 can serve as a graphic reminder of this release, listen, and iterate loop.

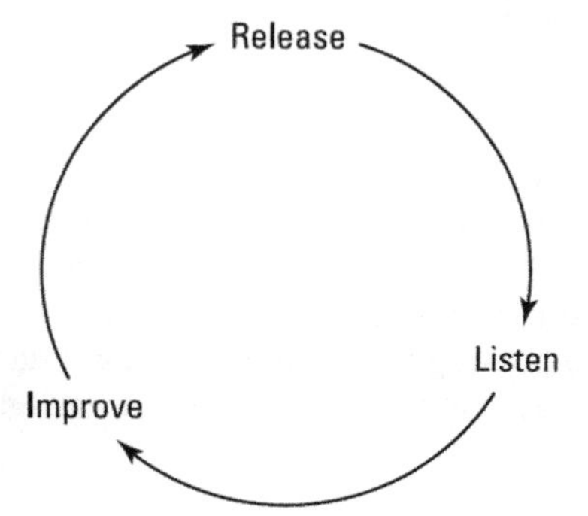

FIGURE 7-3: Testing your hypothesis with customers.

To beta or not to beta?

An alpha release is almost always exclusively to friends and family. These are the people you trust to be kind and not smear you in the media for an absolute boondoggle of a product. They're also the people whom you know (or hope) will be brutally honest with you. Some companies do an alpha release as a final gut check before releasing software to potential customers in a beta.

A beta release might be limited — or it might not. But it creates the expectation with customers that it's not necessarily a polished product. Instead, it's a test. Now, a few of you might be thinking, "Why the heck would I announce that my product isn't finished?"

Well, do you want your customers to trust you? Do you want them to advocate for you? Do you want to attract those early adopters? If so, you have to bring them into the fold. You have to treat them as though they're some of your closest friends. Consumers are smart and won't be fooled by any lipstick you put on the pig. Instead, be brutally truthful. Explain that you've built a thing that you think is helpful, but you're not sure. Ask for help. Give them the space to evaluate your product and give you feedback. Then listen to them. If you don't listen, your customer feedback is absolutely worthless.

Determining Your Customer by Designing a Persona

In DevOps, product requirements exist in large part to bring everyone to the table and establish a shared vision. At the planning stage, you should focus on high-level requirements and trust your engineers to develop the features in the most responsible way given the context provided to them.

The more stakeholders you can involve at the start of the process, the less likely you are to run into unexpected changes down the road. The purpose of gathering requirements isn't necessarily to think of every possible scenario and list every feature you'll ever want to include. Instead, it serves as a way to ensure that everyone involved in the project is aligned with the purpose. When everyone shares one clear direction and a general understanding of the problem that a product should solve, business and engineering stakeholders encounter much less conflict with one another.

Creating a unified vision requires you to fully understand your customer. Who is your customer? What are their problems? How do they interact with products? This is just the start of truly understanding the people who use or will use your products.

Collaboratively developing customer personas is one of the best ways to establish the shared vision of who your user is. In this section, you find out how to design personas to categorize your customers and design features with them in mind.

What is a persona?

The natural inclination of a company is to target its product to everyone. However, "everyone" isn't an audience. Sure, your product might be useful to everyone, but they're not your audience. Instead, your audience is comprised of people who are so excited and enthusiastic about your product that they will evangelize its benefits their friends. To tailor their marketing messaging to these specific users, marketing departments develop profiles of fake people, called *personas.*

TIP

The more detailed you make a persona, the more useful it is in the planning process. In many ways, the persona is a composite of key segments of your audience. It helps you deliver the targeted features and user experience that are the most useful to your customers.

Your persona is a fictional human being with a name, job, background, and preferences. Typically, three to five personas will cover the vast majority of your potential customers. From an engineering perspective, thinking of your audience like this may feel odd. You're used to simply building the products and shipping them. Yes, you're abstractly aware that people use the product, but you're likely not used to thinking about them and their specific preferences before you start coding.

Considering user preferences from the start is why a planning phase exists. It highlights and prioritizes the need to understand the user experience and the values of your customers. Thinking about these aspects from the beginning will inform decisions as you move forward and keep everyone in the organization on the same page.

Designing a persona

Any persona you develop will have a basic profile that includes all the details of their life that are relevant to you. This profile includes basic demographics as well as abstractions such as values, fears, and goals. Your personas should include the basic information in the following list:

- Name
- Job title
- Gender
- Salary
- Location

REMEMBER

I included gender in this list, but I encourage you to consider nonbinary folks and ensure that your database is set up in a way that allows for people who don't identify as male or female. Also, make sure that someone's name can be updated in a user-friendly way. This capability is vitally important to trans people as well as other underrepresented and marginalized groups.

I also encourage you to think about the deeper and more emotional aspects of a person. Although a user's job title and salary help you frame your product, the more human aspects of a person influence decision making. These aspects include education, experience, aspirations, and principles, and these are the key qualities that you want to unlock and understand:

- **Education:** What is your persona's educational background? Are they college educated? If they're developers, do they have a CS degree or did they attend a bootcamp? Education informs what we know but it also influences how we learn — something important in documenting your product and designing the user experience.
- **Goals:** What are their aspirations? These could be professional or personal. Perhaps they want to learn another language or get a promotion. If you can solve a user's problem *and* help them achieve a goal, you will have earned a customer for life.
- **Challenges:** What does this person struggle with? What do they find hard about their job or their life? What do they absolutely hate doing? The more you can unearth someone's pain points, the better positioned you will be to help relieve them.
- **Values:** What principles guide this person? What are they concerned about? What are their politics? You may think that these issues have nothing to do with your product, but they can influence a user's decision to purchase something from you. Perhaps you do business with a company or government that they consider unethical. Think through potential conflicts of interest and how they may impact you and this project.
- **Fears:** Everyone fears something, and most of us carry around many deep-seated fears. Many engineers, for example, fear being made to look stupid. This fear prevents them from asking "dumb" questions which, if answered, could save time and money down the road. If you can discover the fears of your customers, you can address them before they even have to ask. Doing so establishes a tremendous amount of trust.

» Architecting for change

» Documenting design decisions

Chapter 8

Designing Features from a DevOps Perspective

Adopting DevOps is a commitment to infecting every person, process, and product with the core philosophies of DevOps. The software your team produces is in many ways an artifact of the values and principles of your team. If they don't embody the methodology, neither will your technology.

One of the key missteps of a product team is to bring engineering into the design process too late. You've already ensured that everyone at your organization is aware of the product, understands the core business objectives, and has been involved — or made to feel welcome — in the planning process. Your colleagues have collaborated in the brainstorming process, offered suggestions, and come to appreciate which features are most critical to the products success.

Don't let that information sharing stop when the designing of the system begins. Yes, decisions must be made, and sometimes it can feel like there are too many cooks in the kitchen. I'm not suggesting that you hold a democratic vote every time you come to a fork in the road. Hierarchy plays a critical role in most engineering teams, and leadership should be willing to make clear choices when presented with all the options.

But that's the key: *They're presented with all the options.* Presenting all the options requires having everyone involved. The momentum you build throughout the planning phase should be continued through the design phase.

In this chapter, I take you through thinking about software design from a DevOps perspective. I also introduce continuous improvement and show you how to design software flexible enough to adapt to the ever-changing needs of your business.

Constructing Your Design

Architecture, in software, refers to designing the high-level structure of a system. It encompasses not only the actual design but also architectural documentation — the latter being a quality often missing in software systems. Think of it as a blueprint for your entire software product, showing each of the pieces required to make it work and how they relate to each other.

Though you're working with a high-level design, dozens of considerations go into building a solid framework. Handling all these considerations requires more than the skill of a product owner. Strong participation by engineers is fundamental to your success in building a thoroughly developed and maintainable software product.

People have lots of ways to describe "good" code. I prefer to say that well-developed software is performant, meaning that it can perform at the level you need it to in the manner in which you desire. This definition is purposefully vague because it depends heavily on your product and user. Performant can refer to the speed at which your application loads and delivers usability to the user. It can also mean data reliability and availability — for example, is your customer data accurate and accessible at any time? Before you proceed, think through what "performant" code would mean to you, your team, and your customers. How can you design your software to prioritize your specific performance needs?

In traditional engineering teams, developers — those who wrote the code — were separated from operations specialists, that is, the people responsible for deploying and maintaining the infrastructure required to run the application. This separation created what people in tech refer to as the "Wall of Confusion," the proverbial wall over which developers tossed code to operations for deployment and maintenance.

Many organizations — even those who think they've adopted Agile or DevOps — still silo their teams and pass work between each. The business folks decide what product the market needs; sales interjects with features that customers have requested (or been promised); project owners get to work designing the product flow and architecture, only then passing it off to engineering to build.

This type of workflow is an anti-pattern of DevOps (refer to Chapter 9 for more on anti-patterns) and will absolutely corrupt what you are trying to build. Engineers are not code monkeys. Their job isn't to pump out 40 hours' worth of code every week. Instead, their job is to be subject-matter experts, and you will never be able to build a maintainable software product without their input.

Figure 8-1 depicts how engineering teams have traditionally lobbed work at each other in a linear way. Roles were clearly defined, and the space between those roles created friction. Instead of working together to ensure that a feature worked, quality assurance (QA) engineers and developers argued over whose job it was to deal with a bug.

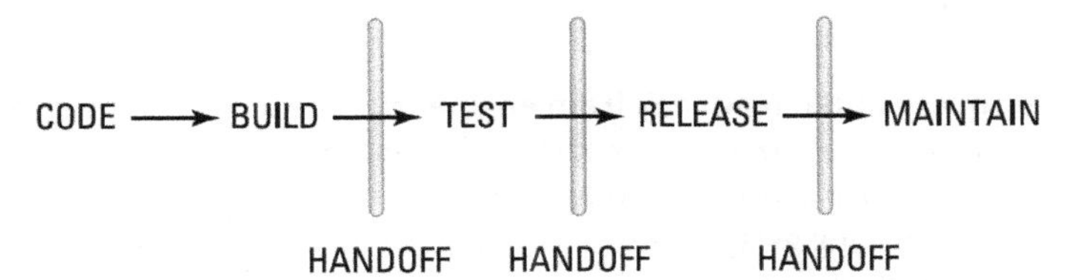

FIGURE 8-1: Handoffs between engineering roles.

With DevOps, your teams should continuously share knowledge and collaborate. Roles refer to a specialization rather than specific boundaries that someone must work within. Previously, roles were essentially air-gapped; no cross-over occurred between roles. DevOps creates an environment of knowledge sharing and a responsibility bleed between each role and stage of software delivery. That way, developers write automated tests for their code and rely on test engineers to find edge cases or suggest improvements.

In this part of the book, I discuss each stage as if it were linear. I take this linear approach to look granularly at the specific ways to inject DevOps into every phase. In Part 3, I connect the circuit and begin to visualize the software delivery life cycle as a truly continuous cycle of improvement.

Unsurprisingly, project owners sometimes come up with ideas that aren't possible or simply aren't feasible within the project's resource constraints. Because of their education and experience, engineers can quickly identify how best to go about solving the problems of your users. They can also evaluate features for complexity and give general advice on the potential costs of time and resources.

WARNING

In my time as an engineer, I've never seen a single estimate of development time turn out to be accurate. The amount of time needed is always an educated guess. The inaccuracy of time estimates is compounded if engineers feel that they must adhere to a specific timeline, even if that expectation is unspoken. I always recommend doubling time estimates to give yourself plenty of room for unexpected roadblocks and troublesome development challenges. Trust me, your customers will not complain if you deliver a product six months ahead of schedule. But deliver late and you'll almost certainly face unhappy users.

The best way to mitigate potential roadblocks later in the development process is to invite everyone to the table at the beginning. This planning and design phase gives you the opportunity to collect everyone's ideas, opinions, and concerns. Although the process of collecting information can seem chaotic, with a cacophony of experts all vying to be heard, the result, when applied, often leads to better software. Here are three specific approaches to keep in mind when designing software:

- **Invite initial feedback from engineers.** Allow engineers to give their insight into how complex each feature will be to build during this discovery and design process. Product managers benefit from the expertise, and engineering gets a sneak peek into what they may be building (and which technologies they need to brush up on).
- **Allow time for analysis.** Schedule time in your engineering team's calendars to ensure that they have the bandwidth to speak with product managers, study the proposed ideas, and think through possible solutions. You might not be able to get all your answers during a three-hour meeting every Wednesday. Most likely, your engineers will have to think through the problems and get back to you. Allow them to do some research to give you the most educated and thorough advice they can.
- **Consider appointing a special team of engineers for the design level.** Some companies are large enough to support the creation of an architecture team. Again, with DevOps in mind, this is not a team that gets tossed a bunch of features and then sorts them out. Instead, they are a specialized group of engineers whose experience or interests make them uniquely qualified to design high-level systems. Instead of diving deep into low-level feature design or infrastructure, they think of the product as a whole — a network of different features that interact in specific ways.

TIP

Creating architecture positions gives your engineering team two paths toward promotion: management and technical expertise. Engineers who don't want to be managers should never feel that they must go that route simply to get a raise or a new title. Career growth for an engineering professional should include a path toward becoming a principal engineer. Systems architects require knowledge from many areas along with the experience to know what works (and what doesn't).

Designing for DevOps

In DevOps organizations, software isn't just built; it's designed. Each part of it is carefully considered and designed to benefit the end user as much as possible. As you and your team work through the capabilities of the software, keep these three principles in mind:

- You should design and build updatable software.
- You should constantly improve your software.
- Your software should support learning.

Designing software for change

The reason most developers loathe working on extremely old codebases isn't that the original engineers were a bunch of idiots. That's far from the truth. It's that the context and circumstances have changed so drastically that the old code is radically different from what it should be. It's antiquated, and that happens at record speed.

Unlike some other industries, software development hasn't existed for hundreds of years, and software developers haven't had the luxury of having the fundamentals already figured out. In many ways, software is still in its infancy. It evolves rapidly, and developers are along for the ride.

Build your software so that it can be updated as the software changes and grows to meet the new demands of your customers. If you need to pivot, ensure that your architecture is flexible enough to endure those changes gracefully.

In addition to designing for change, you need to build your software in ways for it to be reused. Component-Based Development (CBD) is an approach of developing components that are reusable and more easily maintained than large-scale, tightly coupled systems.

Although the code will change a lot in the future, you want to limit the changes you make to your overall architecture because those types of changes will have a more dramatic impact on all areas of your system. Designing a flexible and resilient architecture takes more time upfront but benefits you during acceleration in the later stages of your development life cycle.

To design for change, make the components of your system

- Self-contained
- Independent
- Well-documented
- Standardized
- Portable

Each of these qualities enables code to be lifted up and easily adapted for another location within your software, or to run on a different piece of hardware. These qualities also improve your ability to maintain software over time because the code is easily understood by new engineers on your team. While I'm on the subject of change . . .

Improving software constantly

You need to be continuously improving your software and systems, including checking new designs and expanded code for consistency. Does a given decision fit in with the overall architecture? Is the design standardized to look like other services or components?

The more your system scales and the larger your team becomes, the more likely your codebase is to start looking as though dozens of developers are working on it rather than one. At first, this doesn't sound like such a problem, right? I mean, dozens of developers *are* working on your software. Isn't that just natural? Yes. But natural isn't always good.

Ideally, your codebase (including infrastructure code) should be as uniform as possible. In a perfect world, it would look like one person wrote the entire thing. Now, that level of perfection is unachievable, but that fact doesn't mean that you can't strive for it.

Writers have editorial guidelines. They rely on these style guides to ensure that their language is uniform with their peers. In a similar vein, software engineers have linting tools, which are a godsend for helping their codebase achieve uniformity. Installing linting tools allows your team to decide on basic principles of style, set the configurations, and let the linter fix minor things, such as ensuring that a semicolon appears on every line.

If your team isn't willing to come together and decide on some basic guidelines, well, you've got bigger collaboration problems, and I suggest that you look over Part 1 of this book and start persuading some of these folks to move on to the DevOps way of thinking.

A linter is great for small things. But bigger decisions, like what your base API design should look like, require more forethought and much more discipline. I can't recommend code reviews strongly enough for effective API design. The reviewer doesn't even need to be a senior (though that is often extremely helpful). By reading the code, the reviewer can ask questions and bring up scenarios that the original developer didn't consider. This approach prevents siloed development and keeps major (and possibly mistaken) assumptions from being coded into your system.

TIP

Have your architects or most senior engineers attend all code reviews. I talk more about the code-review process in Chapter 9, but having an architect review new code before it's integrated into the larger codebase will keep the overall architecture standard as well as allow both the architect and the engineer to benefit from the shared knowledge. The architect learns how the feature was developed and by which tools. The engineer learns how to keep their code uniform with their peers and ensure a tidy, well-maintained codebase.

Documenting your software

Well-documented software is incredibly rare, and for a reason. You have deadlines. I've never met a developer whose documentation was tied to their promotions or rewards. Maybe that should change.

Documentation serves multiple purposes, but it should be for more than simply putting what the code says into English. Look at the Ruby method, `add()`, in the code that follows. The function itself is simple enough that the comment is unhelpful.

```
# returns the sum of x and y
def add(x,y)
  x + y
end
```

WARNING

Comments in the code like the preceding example could arguably be useful depending on who has access to your codebase. For example, a JavaScript developer will appreciate some translation of C++ code. The risk, however, is that excessive comments will stop people from reading them because of comment fatigue, or that technical debt will accrue. Documentation must be updated as code changes. Unnecessary commenting can quickly become outdated and lead to more confusion.

Your code needs to be documented in a way that teaches the developer looking at the code about that code. Also, the developer needs to understand not just a particular section but also how that section fits into the whole. What assumptions were made? What were the alternatives? Why was this implementation chosen? What TODO items are still needed? How much technical debt did this one piece of software add?

Transferring knowledge from one developer to another (or the developer to their future self) is the most compelling benefit of documentation, and certainly embodies the values of DevOps.

If you build your software to endure change, continuously improve it, and treat the code as a living document that transfers knowledge about the code and product among engineers, you'll be well positioned to maintain your software over the long term. High-quality design leads to less conflict and faster cycles down the road.

Architecting Code for the Six Capabilities of DevOps

When an engineer and product owner collaborate on the technical design of a new product, the engineer advises on the functional elements as well as on how they interact. The architecture set at the beginning of a project ripples out into the decisions made down the road. The framework determines to what degree the system is flexible for changes and how limited it will be to the addition of certain elements.

Architects influence how the system is structured, which features are prioritized, and how to standardize code. They ensure reusability as well as how the engineering team will tackle the work ahead of them.

Generally, the architects will consider and weigh the six key categories of performance discussed in the following sections.

Maintainability

Code changes. New features must be added, old ones must be deprecated (left functional but unmaintained), and current features must evolve. Software must be upgraded. Change is inevitable and should be planned for. The maintainability of your system is tied to how resilient it is in the face of change.

Your code must be thoroughly tested through an automated test suite. Manual testing gets you only so far. Systems are way too complex these days to have one or two people clicking around a site to see whether everything works. I dive further into testing in Chapter 10, but for now, remember that the code needs to fulfill the acceptance criteria. In other words, how do you know that a feature works? You must also test it for edge cases. You can imagine some. What happens when an array is a parameter but it expects a hash? What if the email address is missing an @ sign? Then, every time a bug arises, you need to add at least one test to verify that the bug is fixed. These tests alert you to when new code has broken current features before customers are affected.

You must document your code. Documenting code is something that almost all engineering teams struggle with, and the problem is rooted in the fact that engineers know they aren't measured by their documentation. They're almost always stressed about getting a feature out or squashing a bug. Those are the measurements they're evaluated against. You can encourage documentation by adding it as an evaluation criteria in your review process. Emphasize documentation in code reviews. Make sure that the decision matrix is documented as well as the end result. What alternatives did the developer consider? Why did they choose this specific implementation over others? That context is extremely valuable for developers (even the ones who originally worked on it!) later down the road. And remember that when code changes, tests and documentation should, too.

Scalability

The scalability of your system is defined as its resiliency when it comes to (sometimes extreme) growth. The best way to think about scalable systems is this: If your system performs better after new resources are added, it is scalable. If not, you may have some work ahead of you.

WARNING

Attempting to scale an application is tricky early in the process. It can trip up startups and products in the early stages because if you don't even have 200 users, your first concern isn't scalability — it's getting more users — any users. Thinking about potential options for the software to withstand growth as you expand is a good idea so that you create the potential to handle accelerated customer adoption and usage. But don't let the goal of growth impede progress on the items that are most critical to your business right now.

One of the advantages of building cloud-native applications — systems designed to run in the cloud — includes their addition of a number of improvements to the capabilities. Resilience, flexibility, and — you guessed it — scalability can be automated and improved iteratively. Whereas manual scaling involves engineers who manage servers, networking, and storage, cloud vendors automate much of this process so that the same configurations are applied to every deployment.

Scalability has an element of elasticity as well. Can your application handle a spike in traffic without falling over? Do your performance metrics stay relatively the same before, during, and after a surge in your application's load?

When evaluating your application's scalability, consider the peak load that your application can currently handle. What impact does an overloaded database have on other areas of the application?

Here are two approaches to scaling infrastructure:

- **Scale up:** You can improve the nodes you already have in use by adding more compute, memory, storage, or network to it. Public cloud providers typically handle this by shifting the application to more powerful instances. From an application perspective, you can play with cache sizes, threading, and increasing connections.
- **Scale out:** Horizontal scaling is most commonly seen in globally distributed systems. It adds nodes of preconfigured infrastructure to your system as needed. If you've ever heard the term "pay as you grow" from a cloud vendor, horizontal scaling is what that term refers to. Scaling out enables you to tailor your scaling to specific geographical regions.

No matter how you choose to design your system to scale, ensure that it fails gracefully (see the "Usability" section, later in this chapter, for what I mean by failing gracefully) if it doesn't scale as expected.

TECHNICAL STUFF

The term *cloud native* refers to building applications with the cloud in mind. It means more than just having an application that is deployed to the cloud. Yes, that is a component. But even applications built before the cloud can be hosted there. Cloud-native applications are developed and deployed with specific cloud-based architecture choices. The teams that build cloud-native applications use a variety of the tools discussed in this book: DevOps, microservices, continuous integration and continuous delivery (CI/CD), and containers. I cover CI/CD in Chapter 11, and containers and microservices in Chapter 20.

Security

DevSecOps was born out of the DevOps movement and exists to remind the community that security is everyone's responsibility. Just as developers and operations folks have traditionally had an adversarial relationship, security has been forgotten by both. Although you can't just interject anything you want into "DevOps," the term *DevSecOps* serves as a good reminder that engineering has many other specializations than just development and operations.

Previously, software was reviewed by security at the very end of the development life cycle. Security had the job of blocking the release of insecure code. As you might imagine, developers don't love hearing that their code is insecure and needs to be fixed only after they've completed it. To address this problem of late notice, the DevOps community has pushed the idea to "move security left." This phrase refers to addressing security concerns earlier in the development life cycle, or if viewed linearly, left on the pipeline.

Securing your software isn't a choice. But securing it at the last minute is too late. It becomes a blocker and reduces your overall flow. With DevOps, you're enabled to bring security into the planning and design process much earlier.

Planning for security

It's important to assess and respond to threats before they become security incidents. Security issues are best researched and evaluated in the planning process. More threats are out there than you know or that you may have the resources to mitigate. But when you come across a threat, you have three choices:

- **Reduce the threat.** Add safeguards into your application and eliminate vulnerabilities. Train your developers to avoid simple security holes like privilege escalation and SQL injection.
- **Transfer the threat.** In some situations, placing the onus of the threat on another organization may make more sense. You can purchase insurance or outsource certain security needs. Still, those actions don't replace the need for basic application security.
- **Accept the threat.** If you evaluate the cost of counteracting the threat and it outweighs the cost of actually dealing with an incident, it might be best to simply accept certain risks as part of doing business in tech.

Security threats

I am not an expert in security, but in case you don't know where to start, here are a few basic security principles to keep in mind when you're designing your system:

- **Privilege escalation:** Bad actors gain access to parts of your system and then escalate their security privileges.
- **Viruses and worms:** Software can be built to replicate itself and infect entire systems. Worms replicate so frequently that they crash the system by taking up too much memory.
- **Ransomware:** This is a type of malware that blocks your access to your own system and holds it hostage.
- **Out-of-date software:** Security holes are regularly patched with updates. Ensure that your systems stay up to date with third-party software updates. If you can't update for some reason, ensure that you know what vulnerabilities exist and attempt to mitigate them in alternative ways.
- **Poor passwords:** Pass good security habits on to your customers. Enforce password rules that make your users choose difficult-to-guess passwords.

You can discover and mitigate security threats at every point in the development life cycle. Be sure to enable your security team to be part of this DevOps journey with you and give them a seat at the table at each stage.

Usability

The concept of usability describes how easily a customer can use your site. Ease of use is at the core of user interface and user experience (UI/UX) design. Any interaction of the user with your application should be designed for usability. In Chapter 7, you see how to plan for the basic flow of your application. You also see how each action of the user leads them to the next and, taken together, these actions encompass the main feature and add value to your service. Now, during the design phase, you ensure that the flow of these actions goes smoothly. Here are some questions you should ask yourself as you design your application:

- **Is it intuitive?** Do users need training to use your application? Do they require previous knowledge? Can a user quickly learn how to interact with the site without much assistance?
- **Is it quick?** Does the site respond in a time frame that is acceptable to the user? Speed performance is important to keep customers from dropping off, but you also don't want the spinning wheel of death — that wheel icon you

see when waiting for an action to complete — to be their main takeaway from your service.

- **Does it fail gracefully?** When an error occurs, what happens? Does the user see a clear error message that explains what went wrong and how to fix it?
- **If a process went smoothly, does the user see a validation message?** Communicating with the customer via messages is a way of failing gracefully. In other words, the application encountered an error and passed the error along to the customer instead of simply crashing. I discuss engineering for error in Chapter 10 and failing well in Chapter 16.

TECHNICAL STUFF

Not sure where to start with usability? Evaluate your signup process first. It's kind of like cleaning the bathroom if you're short on time and have guests coming over. That's the one room you can be pretty sure everyone's going to see. If your signup process is usable, you'll be able to iteratively improve on the usability of other aspects of your site through various tracking tools. But if they never sign up, you won't know whether your product is a flop or your site simply wasn't user friendly.

Reliability

The reliability of your system comes down to the availability of your software to users. This reliability includes the accuracy and integrity of data stored in your database as well as what's visible to the user. If data becomes inaccurate or out of sync, the system is not reliable.

If the system does go down, how easy is it to restore? What is your mean time between failure (MTBF)? (I cover MTBF in Chapter 17.) What expectations of availability do your customers have? These expectations could be assumed or legally binding via a service-level agreement (SLA). Data inconsistency can become a problem if backup data is used to restore a system after an incident. How do you ensure consistency in those situations? What are your redundancies?

Here are some terms to keep in mind when planning and evaluating your software for reliability:

- **Availability:** The percentage of time your system is functioning and accessible by customers.
- **Latency:** The time between when a user makes a request and your application responds.
- **Throughput:** How many transactions an application can manage per second.

- **Fidelity:** The level to which your application represents the actual state of an object.
- **Durability:** Your application's capability to meet the expectations of your customer over the long term.

Flexibility

A flexible system is one that is the most capable of evolving to meet the needs of the customer. Flexible codebases can absorb new code without the possibility of major disruptions. Here are questions to ask yourself and your team as you're designing for flexibility:

- If you're using a SQL database, can the scheme accommodate change well? How difficult will updates be?
- What does your dependency tree (the visualization of tools or other software that a piece of code depends on to run) look like? Which services are vulnerable to chained failures because of dependencies?

TIP

  The term *chained failures* refers to the impacts on your application based on failures "upstream" in tools on which you depend. For example, if AWS experiences an outage, and your application is hosted in AWS, your application experiences an outage as well.

- How easily can new components be integrated into the overall system? How do components communicate?

Documenting Design Decisions

I touch on documentation in some previous sections in this chapter because documenting your process is so important to the planning portion of the development life cycle. Teams often make decisions at the beginning with the benefit of all the information and then forget that their future selves won't benefit from the same context.

Creating great architecture is not enough. You must go one step further. Document the alternatives you considered, the costs of the path you chose, and the reasons you made the decisions you did. If you don't write down these aspects, that knowledge will be lost. You will not remember it — I promise you. And even if you're lucky enough to possess eidetic memory, that knowledge should not be kept in your head. You're adopting DevOps and must share information with your peers.

If you give your team all the tools you used to make the architecture decisions and design the system, you enable them to reuse the design. Take the time to write it down. Even disorganized notes are more useful to your team than nothing at all.

You could use a documentation tool to store your thoughts, but I recommend a different approach. Store your design decisions — and the thought process behind them — with your code. Yep, right in your codebase. Create a markdown file in the root directory titled "Architecture Decisions" and do a brain dump.

The impact of the decisions you make early in the process ripple out. Those decisions impact every part of the system, from the code to the infrastructure. Those components and the things that link them are intertwined. If you want your team to maintain this design, or be empowered to change it with all the context needed, please clarify everything they need to know. Otherwise, your architecture will drift away from the original design and unnecessary complexity will begin to strangle your system.

Avoiding Architecture Pitfalls

Keep in mind the following basic architecture fundamentals that will carry you and your team through this design phase of your development life cycle. Each suggestion isn't necessarily rooted in DevOps. Instead, all the ideas support the DevOps philosophy and enable your team to collaborate more fluidly, take responsibility for the quality of system as a whole, and develop better software faster.

- **Understand your full stack.** People in the industry use the term *full-stack engineer* a lot and never seem to quite agree on what it means. I've met only a handful of people I would describe as full-stack engineers — that is, people who understand the system from the hardware to the operating system to the language and frameworks used.
- **Isolate components.** If you adopt CBD or build system microservices, ensure that they are scalable and modular. Reduce or eliminate shared state and prevent accidentally coupling microservices into what I like to call "macroservices." The last thing you want is to have all the downsides of microservices (see Chapter 20 for more about microservice architecture) with none of the benefits.
- **Don't make difficult choices configurable.** Future you — and others — will choose wrong. Take the time to gather your senior and principal engineers, as well as relevant subject-matter experts as needed, and decide on the best course of action for every scenario you can think of. Ideally, you will make

choices automatically for the developer; using a configuration setting (and suggested default) can be a second option. The more choices you can remove from a developer as they're coding, the less likely they are to make poor choices.

- **Document configurations.** Always include a default and add a few examples to help the engineer understand the potential impacts of their choices. This approach improves uniformity, reduces human error, and teaches engineers more about areas of your system with which they may not be familiar.
- **Keep your system dynamic.** Avoid developing your software for a specific ecosystem or tool whenever possible. This type of vendor lock-in is dangerous to your long-term outlook because it makes transferring to better tools down the road difficult. It also impedes your ability to make changes and evolve. Generic and stateless components are the most flexible and can be picked up out of one environment and run in an entirely different one.
- **Use a log aggregator.** Don't log directly to the file system. In the event of a crash, restoring valuable information from the logs that will help you remedy the issue will be impossible. The same applies to VMs and containers that are destroyed. With an aggregator, the logs outlive the nodes and provide you with opportunities to customize how much information is stored.
- **Avoid calling infrastructure APIs from your app.** Calling from your app makes switching infrastructure tools or hosting providers more difficult. Instead, look into open source Platform as a Service (PaaS) tools or vendor products that can help you abstract the infrastructure from the application running on it.
- **Go with the crowd.** There's a time for building your own tool from scratch, but it's extremely rare. Don't use obscure, self-made tools or protocols unless you absolutely have to. Take advantage of the tools and protocols used by thousands of developers every day. This suggestion applies to everything from HTTPS to standard database connections to REST-based APIs. These are the tools that provide the most capability out of the box as well as have the best documentation around the tool. That documentation includes questions that are posted to forums like Stack Overflow, where you can get answers from Google quickly rather than think through a simple problem for hours.

» Writing maintainable code

» Making decisions with DevOps

» Implementing good practices

Chapter 9

Developing Code

I wrote this chapter with operations folks in mind. I hope to empower them and managers who don't have an engineering background to understand the process of developing software — and the hundreds of decisions it requires. It will enable operations people to feel more confident discussing code and increase their empathy for the decisions (and accompanying mistakes) that developers make daily.

If you're a developer, a lot of the content in this chapter may feel familiar to you (though I might argue that you could still use a refresher on good development practices). Chapter 11 goes into depth on releasing code, choosing a deployment style, and versioning — topics about which most developers are less confident.

In this chapter, I show you how to talk about code in a collaborative way, write code that is agile in the face of change, and make software decisions from a DevOps perspective.

Communicating about Code

The caricature of the basement-dwelling hacker is antiquated. Although sufficient diversity and inclusion in tech remain a challenge, the situation has improved. More and more people from nontraditional backgrounds have joined and brought their diverse education and experience to the industry. One of the benefits of this diversity has been the emphasis on communicating about code, which is wholly in line with the values of DevOps.

I discuss code reviews — the process of having a peer review your code prior to merging into the master branch — later in this chapter, in "Having peers review your code." But communicating about code starts much earlier than the code review stage. Developers used to be handed requirements and expected to develop the appropriate features, only to hand the code off to QA and security for review and operations for deployment. DevOps has changed all of that.

Today, developer communication is critical to the acceleration required to differentiate your business from competitors. Engineers on the development side work closely with different areas of the business to understand the context of a feature or product before requirements are set and user stories are created.

A user story is an Agile approach to describing a feature from the user perspective. Traditionally, you would have been lucky to get vague requirements like "Create user signup process. Require email and password." Instead of creating enormous tasks with vague requirements, user stories give the developer specific detail from the viewpoint of the end user and break large features into small pieces. Here's an example of a user story: "As a site visitor, I want to click a link on the home page and be directed to the signup form." That story could be followed by, "As a site visitor, I want to fill out a signup form with my email and password, click Submit, and receive a verification message that my account was created for me."

If your team doesn't communicate well, you need to take time to implement some of the practices that influence good communication. Code reviews and post-incident reviews provide the opportunity to practice communicating as a team. As I say elsewhere in the book, communication is a skill just like any other. It can be learned, but it takes time to master.

Give your team the tools they need to improve in the "soft skills" needed to be a great developer. Speech coaches and improv classes can radically improve the skills of someone who struggles to communicate. The fact is that most people could benefit from some kind of coaching on how to relate with others and show more empathy to their team.

DevOps brings all stakeholders together, and communication is a critical component of that goal. If you find that your development team is homogenous, and you have room for increased head count, breathe life into it by hiring some developers who bring different viewpoints to the team.

I highly discourage you from hiring one woman or one person of color onto a pale and male development team — especially if that person is a junior developer. People from underrepresented and marginalized groups do much better when they have the ability to vent and amplify the voices like their own. A single developer who represents a specific group is likely to be discriminated against, sidelined, and gaslighted by the rest of the team. I've been that person. It's an extremely difficult position to be in and one in which that person is unlikely to thrive. Having the one "other" in the group as a junior only reinforces old stereotypes that certain groups are less qualified or talented in engineering. DevOps is an extremely inclusive community, and for good reason: It's the best way to build great products. Make sure that you emphasize the same on your teams.

IMPOSTER SYNDROME

When I work with teams looking to adopt DevOps methodologies into their everyday work, I often find that the developers on the team feel a little self-conscious about their lack of operations knowledge. Likewise, many operations folks feel self-conscious because they don't know how to develop software from scratch. Even as they're learning, both sides can suffer from some degree of *imposter syndrome,* which describes high-achieving individuals, like you, who struggle to internalize their accomplishments and experience a persistent fear of being exposed as a fraud. I struggle with this fear, and many people in the tech industry struggle with a feeling of being less than — of not producing fast enough or not working hard enough.

Imposter syndrome can impact your ability to create a DevOps learning environment in a couple of ways:

- **It makes you feel less than.** If you feel less knowledgeable or talented than your colleagues, you will be less likely to ask questions that you think might make you look "dumb." Not asking questions is the absolute worst thing you can do because it cheats you of learning and your colleague of teaching. Further, it compounds the unspoken fear of asking questions across your team.
- **It cheats you of the confidence to teach.** You know more than you think you do. You also have much more to contribute to your team than you currently feel confident in doing. Imposter syndrome is that tiny whisper that says, "You're not the expert." (So what if you're not?) Just because something is "easy" to you doesn't mean it's not hard for your colleagues.

In a perfect DevOps culture, engineers will fearlessly embrace what they can teach and openly receive what they need to learn. Until you're there, the traditional friction between developers and operations people can surface in the development phase of software delivery because it's the developer's domain. At the development stage, developers feel the most confident and the operations people feel the most vulnerable, and each side's imposter syndrome and pride can stifle collaboration. The truth is that both sides are an absolute necessity to each other; one without the other would be lost.

The more your team practices communicating the way they think about code and write software, the better work they'll produce. The team starts to understand how each person goes about solving problems, questioning assumptions, and building code that is human readable rather than simply machine ingestible.

Engineering for Error

Error handling is an important part of writing maintainable code. Silent errors are one of the most dangerous things lurking in your codebase.

Programming graceful exits can sometimes make code more verbose but will allow the code to handle an error by printing out a descriptive message rather than just quitting or slowing down significantly. Programs without proper error handling often display strange, unexpected behavior that is difficult to debug.

TECHNICAL STUFF

Part of handling errors is ensuring that messages to customers make sense. A 418 status code and an obscure message about a null pointer doesn't assist the average customer, and even the technical ones will likely roll their eyes at you. Build your user interface (UI) to display messages that help the customer understand what went wrong, where to go next, or whom to call for help.

Aside from providing a clear message that allows the developer to understand both what happened and where in the code it failed, ensure that any data impacted is recoverable and consistent. You can't get by with writing programs that work only when everything happens as expected. Great developers think through potential exceptions and edge cases that enable them to write code to handle those conditions. (The term *edge case* refers to an improbable but possible scenario.)

Writing Maintainable Code

You don't write code to run for a day. (Although I often think if an apocalypse somehow spared the Internet, most of your systems would fall over within days.) The software your team writes will most likely be run for years — a particularly daunting thought for anyone who's been embarrassed by code they wrote a few months ago.

Maintainable code is never in its final state. It's alive! (But hopefully it's in better health than Frankenstein's monster). Like the U.S. Constitution, code is, metaphorically, a living, breathing document — and it requires care and forethought.

Testing code

I cover different types of code tests in Chapter 10, so the main point for now is that you should get in the habit of writing testable code. For software to be testable, it needs to be modularized into small components and functions. If *x* is expected to do *y*, you can write a test to ensure that *x* actually does *y*.

Legacy codebases (sometimes referred to as the "brownfield") often don't have tests, or testing is sparse. One of the challenges of maintaining these older systems is that even if you wanted to write tests, the code isn't designed in a way that easily enables you to do that. If this is your situation, you don't have to flush the whole thing down the toilet and start over. Instead, think of it as an old car. You don't replace parts that are working. When something breaks, fix that part and add a test to ensure that the fix is stable.

Debugging code

Debuggers are key to seeing what's going on in your code in (almost) real time. As you may know, debugging tools freeze your program at a specific point that you choose, which is a great way to uncover unexpected results and see what's different from what you expected. For instance, the value of a variable could have been mutated unexpectedly or the wrong type passed in by accident.

The example code that follows demonstrates how a debugging tool or debugging statement (shown in bold) is inserted into the middle a function so that developers can check their assumptions and understand what's happening while the program is running. This example comes from "The Little Guide of Linked List in JavaScript," by Germán Cutraro (`https://hackernoon.com/the-little-guide-of-linked-list-in-javascript-9daf89b63b54`) (Don't be overly concerned with the functionality.)

```
LinkedList.prototype.addToTail = function(value) {
  const newNode = new Node(value, null, this.tail);
  if (this.tail) {

    // insert debugging tool or console.log() statement here

    this.tail.next = newNode;
  }
  else {
    this.head = newNode;
  }
  this.tail = newNode;
}
```

Most IDEs (integrated development environment) and browsers have debugging tools out of the box. Debuggers can be extremely useful for less experienced engineers even when no bug exists. The debugger permits you to "step through" the program so that you begin to think more like a machine and become quicker at reading code.

Logging code

Logging can be a developer's most valuable tool or their worst nightmare. Where debuggers become obsolete, logging provides answers. You can't always step through code at runtime. Instead, your code may be distributed or deployed to the cloud.

Logging is like debugging but instead of putting a breakpoint into the code, you add logging statements that you can read through as a program runs. The logs display the actions and state of the program.

Logging frameworks are tools that classify log messages and help you comb through the logs quicker than you could if the code was simply outputted as raw data. Logging isn't free, though. You have to store the logs somewhere, so you need to log data based on what you need to know. Logging everything would be both a poor use of resources and overwhelming to consume.

What you log, how often you log it, and how you organize it is up to you and highly dependent on your application. Here are three guidelines that I recommend you implement:

- **Format your logged messages.** Include pertinent information such as the session ID or user account information as well as the time stamp and message.
- **Provide context.** Sometimes you need more than the immediate data. Simply knowing that something went wrong is not enough. What activity happened before an error was encountered? What data was impacted?
- **Avoid side effects.** Your logging should not impact your application's performance. Logging everything is tempting but comes at a cost. Instead, start slow. You'll find that you can more easily add logging than remove it after it's in place.

Writing immutable code

One of the biggest benefits to functional programming (discussed in the upcoming "Programming Patterns" section) is its emphasis on immutable code. Basically, all variables are assigned once and do not change. In case threading is a concern, immutability creates more resilient code. Also, the software is much easier to debug because variables don't change in the middle of the program. Instead, a new value is assigned to a new variable. The fewer moving parts you can put into your code, the easier it will be to debug and maintain.

Creating readable code

Your application's code must be readable by the machines on which it runs. But the machines don't maintain it. Instead, humans have to read it, parse meaning, and make changes that won't cause a black hole.

When you think about writing code to be readable by humans, you should consider more than just your colleagues. You should also consider future "you." You won't have the context you have now in six months when you try to unravel why something was stored in an array.

Also, the more legible your code is to humans, the less trouble they'll have making changes and fixing bugs. Sometimes it's fun to make code so concise that it takes up only a few lines. But if your code takes someone else hours to deduce what is actually happening, the maintenance cost is too high.

Programming Patterns

Many more programming paradigms exist than the two I cover in this section, which are object-oriented programming (OOP) and functional programming. Both of these paradigms are simply two approaches to the same end, which is to organize logic into a software program that provides utility to the end user. I choose to highlight OOP and functional programming because they're both popular and give you a wider view of possible approaches because of their contrasting features.

Object-oriented programming

Object-oriented programming (OOP) is based on the concept of — you guessed it! — objects. Objects are anything, really, but usually contain data. Objects may have attributes or associated qualities. In OOP, people typically refer to procedures as functions or methods. Most object-oriented languages — Java, C++, Python, JavaScript, Ruby, and Scala — are class based. Objects are instances of classes.

The goals of object-oriented development are reusability and modularity. Keeping pieces of logic small and with other associated objects and methods is ideal. You can easily reuse functions that have been developed within an object-oriented program, which aids in efficiency and enables you to recycle work already done. This capability for reuse can, however, lead to problems if the developer is undisciplined about ensuring that the method is actually reusable in an intuitive and flexible way.

Object-oriented programs encapsulate logic in such a way that an object does not need to know the details of its implementation for it to be used. Objects can hide certain attributes from programmers, which prevents the visibility of values that no one should tamper with. This approach provides design benefits that reduce the burden of maintaining large programs via relatively easy modifications.

Functional programming

The functional approach to programming avoids changes to state and emphasizes the immutability of data. The output of a function in functional programming may be impacted only by the arguments passed into a method. This approach has no side effects. If you call a function with the same parameters a thousand times, it will always produce the same result. Side effects are avoided because these functions cannot be influenced by local or global state that would impact the result.

Functional programming is extremely modular and easy to test. Its practice allows the engineer to make fewer decisions about writing clean code than is possible with OOP. Clean code is so ingrained into the principles of avoiding side effects and preventing mutable state that functions end up being written in a clean and readable fashion. Additionally, the code has fewer moving parts, which makes identifying where a bug might be relatively simple.

Functional programming was born out of lambda calculus, but developers don't need to be math geniuses to write functional code. Though you don't need to write code in a functional language to benefit from the practices, functional languages include Lisp, Haskell, Scala, Erlang, Rust, and Elm.

Choosing a Language

Choosing the right language for any project is a difficult decision. You have countless options, and each one has its pros and cons. Also difficult is knowing how to separate hype from genuine praise to determine which language will give you the best tools for the job.

No single language is superior to all others, no matter what the evangelists of any particular community may tell you. Each one always has trade-offs to consider.

I can't list every language and its potential benefits (and costs) to your team, but here are some aspects to consider:

- **Performance:** Will the language be performant in the way you need it to be? Benchmarks are available to give you an idea of a language's performance, but keep in mind that the quality of the code also impacts performance. A well-developed Ruby application outperforms a poorly executed Java application no matter what the language benchmarks are.
- **Comfort:** Does the team know the language already, or will they be able to pick it up quickly?
- **Community:** Can you easily find answers to questions online and locate community resources formed around the language?
- **Platform:** Does the language require a specific machine or tool? For instance, programs developed in Java may be run only on machines with a Java Virtual Machine (JVM) running.
- **Framework:** Some languages are tied heavily to the framework. Ruby is a perfectly useful language on its own, and lightweight frameworks like Sinatra exist, but Rails is married to Ruby in many ways. Think through how that fact will impact development.

If you opt for a microservice architecture and have a large enough team, you might be able to build your application using multiple languages. Each service can interact with services written in another language through a standard protocol or API.

Avoiding Anti-Patterns

Anti-patterns describe behavior in software development that highlights poor practices. Anti-patterns frequently appear to make sense at first glance and often seem to be commonly practiced in the industry. The consequences can be severe, however, and other solutions have proved more effective. Although many more anti-patterns exist than those in the following list, here are the software engineering anti-patterns to avoid in your DevOps practice:

- **Design by committee:** Because of its emphasis on collaboration and communication, sometimes people can interpret DevOps as being a design-by-committee pattern of software development. It is not. That type of decision-making results in horrible outcomes. Instead, come to the table having already thought through the process as individuals. When multiple parties come together to share the ideas they've individually thought of and then discuss them, the outcomes are vastly different than when a group of people get together without any forethought and must come to a consensus.
- **God objects:** This anti-pattern surfaces when too much logic is contained in a single part of your application. This omnipotent object or class wields too much power and forces other objects to rely on it. Maintenance becomes difficult because the code becomes so tightly coupled and the god object so large that the code is difficult to debug.
- **Cargo culting:** This term refers to implementing a specific pattern of development or tool without understanding whether or why it's the best solution. Though the pattern or tool is most likely implemented by a more inexperienced developer, even senior engineers implement a cargo cult solution if they're influenced by vogue tools or constrained by tight deadlines.
- **Law of the hammer:** If your engineers rely too heavily on a language, framework, or tool that they're intimately familiar with, they may be suffering from this anti-pattern. Your engineers should be comfortable with their tools, but if comfort becomes complacency, the time has come to reevaluate whether you're using the best tools for the job.
- **Bleeding edge:** This term describes engineering teams who opt to be early adopters of technology and integrate it into their applications. These new technologies, although novel and occasionally amazing, can be unreliable, poorly documented, and buggy. You also risk using a technology that's incomplete or a beta that pivots hard before release and thus impacts your code.
- **Overengineering:** Any time you're designing a product, you must discipline your team to solve only the problem at hand and to do so in an efficient manner. Making a process unnecessarily complex is overengineering. Although overengineered safety functions are necessary when lives are at stake, this scenario is rare and should be avoided by most developers building products. Keep it simple.

- **Spaghetti code:** This term refers to any object or application whose code is unstructured to the point of being barely readable. The code may function (barely) but it's twisted like spaghetti on a plate.
- **Copypasta:** This anti-pattern is simply copying and pasting existing code — or code you found on the Internet — into your application. If this is a solution, create a generic solution that can take parameters for customized handling.
- **Premature optimization:** Engineers can be tempted to make something as efficient as possible right from the start. But optimizing prematurely is often not the best use of resources and can make code more difficult to maintain — especially if you're not completely sure that you've solved the problem. MVPs should never be optimized, and optimizations should take place only after they've been identified as necessary.
- **Vendor lock-in:** I mention this issue several times in this book. A lock-in situation occurs when switching vendors would cost so much that it becomes a barrier to opting for a new, and perhaps better, tool.

DevOpsing Development

No one is in charge of your career but you. Sometimes managers fail you and peers disappoint you, but when you come up against disappointments at work, you shouldn't let it sidetrack you from your mission. DevOps requires collaboration, but you have no guarantee that collaboration will always be pleasant.

Being excellent at what you do is a choice, and I believe that hard work beats talent when talent doesn't work hard. When it comes to developers, possessing certain key characteristics makes them both excellent engineers and exceptional DevOps practitioners.

DevOps can't exist on an engineering team without the buy-in of developers, and developers are some of the people who can benefit the most from the DevOps approach. When hiring developers, keep the characteristics described in the following sections in mind. A developer's attitude about their work is as important as their technical expertise.

Writing clean code

Clean code is human readable and simple to test. Each function (or method, in some languages) should do only one thing. This single-responsibility principle modularizes your code so that you can quickly deduce what a function does and where a bug might be.

Functions that lack focus create difficulties in reading the code and fully understanding what purpose a section serves. The lack of focus also makes reusing the logic or abstracting into a generic method for use in multiple areas of the codebase difficult. Ensure that functions are named for what they do. If you catch yourself adding "and" to a function name, take that as a sure sign that the function is breaking the principle of single responsibility.

Understanding the business

An anti-pattern that I didn't mention earlier in this chapter is mushroom management, which describes blind development in which developers are given limited information and expected to develop based on manager decisions alone. The name comes from how mushrooms are grown. Mushrooms are kept in the dark and occasionally fed some manure. In mushroom management, no collective understanding exists of the reason behind a product. The situation is made worse by the fact that managers and developers often have trouble communicating.

If developers don't fully understand the business, they fail to write code in a way that fully serves the right purpose. Conversely, developers who have a handle on the business side feel empowered to suggest alternatives, push back on ideas, and take pride in their work.

Listening to others

In business and engineering, the art of listening is perhaps the most underrated skill of all — especially for developers. If you watch highly productive teams interact, you often find that the senior and principal engineers do the least amount of speaking. In fact, the best technical leaders on engineering teams allow everyone else to contribute their thoughts, consider everything carefully, and then give clear guidance on how the team should execute a plan.

A characteristic I look for in hiring is a person's comfort level with admitting what they don't know. Engineers who think they're the smartest person in the room can absolutely destroy collaboration. They will silence their colleagues and steamroll anyone who disagrees with them. The cost to the team is too high to employ engineers who cannot admit when they're wrong or listen to the ideas of their peers.

Focusing on the right things

I almost never use the word "coder" because it implies someone who mindlessly types code without thinking through larger implications. Possessing the discipline and gumption to push back on ideas coming from different areas of the business in a way that invites discussion is critical. These capabilities require developers to translate technical language into words that non-engineers can understand.

Developers who focus on the right things almost never sacrifice the quality of their work for unreasonable deadlines. Instead, they communicate hiccups early and keep everyone informed of the deadlines for when work is expected to be done (with an emphasis on *expected*).

These types of engineers are cautious about taking on technical debt and quick to pay it off. They focus on architecting and building features that are important to the business, are maintainable, and are implemented in a way that makes the codebase flexible to change. They avoid rabbit holes by keeping the customer in mind and avoiding overengineered solutions.

Getting comfortable with being uncomfortable

Curiosity is a characteristic of all great developers. They aren't afraid of new things and embrace new ideas with a childlike joy. Great developers recognize that new industry tools or trends aren't always the best idea for any particular company, but they keep up with tech news and learn the basics of new tools so that they can make good decisions about those technologies and trends.

Ongoing education is another key component of teams that produce great software. They emphasize the importance of reading, talking to other developers, going to conferences, and taking courses. If you are a manager, be sure to advocate for your developers and block off part of the budget for continuing education. Your developers are more than a way to pump out code. They're a knowledge resource that, if cultivated, can provide years of valuable advice and guidance to your company.

In addition to providing your engineers with educational opportunities, ensure that they have quiet time to develop. An engineering manager I know allows meetings to take place only on Mondays and Fridays. Instead of scattering meetings throughout the week and taking developers away from focused work, he protects them from interruption. Developing software takes intense focus. A single break in that focus can sideline a developer for hours.

Establishing Good Practices

Now that you know what *not* to do, you can focus on how to implement good practices in your organization. And, no, I didn't say "best" practices. A *best practice* is an approach that is viewed as superior across the industry because it produces better results than any other technique. In other words, it's the accepted way of doing things.

TIP

I don't like the "best practices" approach because it stifles innovation. If you accept that something is the best practice, will you challenge it or iterate on it? On the other hand, good practices are standard methods of approaching certain challenges that are generally accepted as battle tested. Good practices give you guidance without imposing rigid constraints.

Organizing your source code

Every engineer on your team should have, at a minimum, read-only access to every line of code in your organization. This access includes the source code of your application's features all the way to infrastructure code. This shared repository (or, more likely, repositories) enables everyone to feel empowered to find their own answers and read parts of the application that they're not necessarily intimately familiar with. With this shared access, every engineer can be useful during day-to-day work and, most important, during incidents and outages.

For most organizations, git and a hosting service like GitHub or GitLab is ideal. These tools are much lighter weight than older source control tools and serve as great collaboration tools — even for meeting agendas and brainstorming!

Be sure to keep related code together. Builds should be simple and repeatable. Also, as you advance, automate your builds as you move toward continuous integration.

Writing tests

If you don't have a testing framework already set up, do that now. Giving your developers the ability to write automated tests as they write features is imperative. Some people opt for test-driven development (TDD), in which you write a test that confirms the function you need to write and then you make code pass the test. This approach is effective but heavy-handed enough that many avoid it. At a minimum, developers need to write unit tests that confirm that a piece of logic performs as expected.

You can use happy path tests, which are scenarios in which everything goes as expected. You can also use sad path tests, which are scenarios into which something odd is introduced.

The automated testing framework that you use will depend on your language. Find one that is robust enough to meet the needs of your organization but simple to learn and execute. If you make testing difficult, your developers won't do it.

Following is code from two example files: `add.js` and `testAdd.js`. The only function in `add.js` takes two parameters and returns the sum. The test file `testAdd.js` accompanies this piece of logic and contains two tests — one with a happy path and one with a sad path. The happy path test provides two expected parameters: 2 and 2, which returns 4. The sad path test introduces a string of 2 as one of the parameters. Although this result is not expected, it is possible, and your logic must account for it.

```
// add.js
function add(x, y) {
  return x + y;
}


// testAdd.js
const assert = require('assert');

// happy path test
it('correctly calculates 2 plus 2', () => {
  assert.equal(add(2, 2), 4);
});
```

```
// sad path test
it('correctly calculates 2 plus 2', () => {
  assert.equal(add('2', 2), 4);
});
```

The preceding sad path test will fail because of how JavaScript attempts to help you handle strings. Adding a string of 2 to an integer of 2 will result in a string of 22.

Documenting features

Making notes above a piece of code is a way of reminding future developers what the function does (if it's necessarily complex), what the context of the code is, what parameters it expects or produces, and what, if anything, could be improved with more time. (Hey, sometimes you have to do things in a hurry.)

The code itself should be clean and readable enough to serve as a type of documentation even though you're writing it in a machine language, not a human language. Just as with everything else in DevOps, you can and should automate documentation — to a point. Just remember to manually solve your problems before you automate them. Otherwise, you'll be automating broken systems. If you do choose to automate your documentation, create the framework and allow developers to configure specific values to tailor the documentation to the specific code.

When I write APIs, I have a script that loads the boilerplate API format with the actions I know I'll probably need (`GET`, `POST`, `PATCH`, `DELETE`) as well as the basic documentation (including examples) for each action. That way, I don't have to type the same things repeatedly. I save time and know that I'm not making (as many) mistakes. Then I take the boilerplate and add to it or adjust it as needed, based on the specific code I wrote. Getting into the habit of automating small pieces of redundant work is a very DevOps-y thing to do!

Another type of documentation is external and customer facing. That documentation typically isn't managed by developers because it requires much more verbose technical writing and assists engineers with getting up and running. As someone who works in developer relations, some of the work I do is showcasing the APIs made by the product engineering team into documentation and tutorials that anyone can understand and use.

Having peers review your code

I believe strongly in code reviews. I also believe that developers should never merge their changes into the master branch themselves. A code review can take place through comments in the repository where the code lives or in person with two (or more!) engineers reviewing the code together on the screen.

The practice of reviewing code is important on many levels because it

- Helps junior engineers level up more quickly
- Reduces errors by having more than one pair of eyes look it over
- Unifies the codebase by standardizing formatting
- Forces reviewing engineers to question assumptions and ask questions
- Enables people to become familiar with code they didn't write
- Helps senior engineers (who sometimes code quite a bit less) stay in touch with how the less experienced think

The process of a code review is simple. Assuming that you utilize git as I recommend, your code will live on a feature branch while it's a work in progress. You will then submit a pull request (PR) to merge your code from the feature branch into the master, or trunk, branch. (Depending on your deployment approach, the master branch may or may not be the version currently running in production, but it will be the most up-to-date version running in the development environment.)

You should tag a particular party in the PR. If you're on GitHub, you can simply include `@username` in the comment of the PR, thereby sending an alert to the other engineer. How you organize who reviews what is up to you. Some companies assign particular people to a team; others leave it more ad hoc.

TIP

If time is limited for you and your team, you can still benefit from even a lightweight code review in which both engineers quickly discuss the purpose and glance over the code. They'll still find plenty of bugs.

If it's a remote or asynchronous code review, the reviewer will look at the code and respond to the PR with any comments or concerns. If you opt to review the code in person (or remotely through a video chat), find a quiet space to review the code as a dyad or small group without interruption. If you are in person, use a large monitor to aid you in easily reading and discussing the proposed code. At this point, the original developer and the reviewer(s) read the code and ensure that it follows your team's code standards, functions as expected, is written in a readable manner, and is properly tested.

After the reviewer is confident that the code is ready to be merged, they merge the code into the master branch. This shared ownership encourages everyone on the team to work collaboratively and treat the entire codebase, instead of only the code they worked on, as their responsibility.

IN THIS CHAPTER

» Using automated testing

» Looking at the various testing environments

» Understanding what types of tests to do

Chapter 10
Automating Tests Prior to Release

Testing and development overlap to some degree because developers should absolutely be writing tests as they write code. I gave the subject of testing its own chapter to highlight just how important testing is to DevOps environments. You can't have automation or continuous anything without robust automated testing.

In this chapter, you glean the importance of testing in DevOps, see how to test code in multiple environments, and find out what types of tests to consider.

Testing Isn't Optional

If you jump into continuous integration or delivery without taking the time to establish a strong automated testing practice on your team, you face disaster. Things will break frequently and catastrophically. Testing buttresses your ability to automate and reassures you that new changes don't break existing functionality.

Software testing has three core purposes:

- **To confirm that application logic fulfills its desired functionality:** Does the current functionality meet requirements and complete the task in a reasonable time?
- **To discover bugs — errors — in code:** Does the logic respond to all types of inputs? Is the code usable by your customers?
- **To verify that previous functionality is unchanged by new code:** Has anything been accidentally impacted due to unforeseen dependencies?

Automating Your Testing

Manual testing is becoming obsolete. Our systems and codebases are simply too complex and run in too many different types of environments for a human to confirm that everything works as expected. If you're adopting DevOps and all its associated practices, automated testing isn't a choice; it's the next step.

Continuous integration requires an automated test suite that runs tests every time code is committed to git. This approach requires not only that your team writes tests but also that you treat your test code as code.

Automation is key to enabling a "shift-left" mentality similar to the one I talk about in Chapter 6. Done well, testing allows you to fail early and often. You catch more bugs, avoid regressive functionality, and prevent incidents in production through continually testing your system.

Manually testing each change is labor intensive and inefficient. You should shift the QA team's efforts from running tests and — face it — clicking around the site manually to developing automated tests. If you're lucky enough to have dedicated testers, treat them as testing specialists. They are the experts in the best testing frameworks and tools, as well as how to automate the test suite for accuracy and performance. Developers should absolutely always write tests to accompany their code. Similar to a code review, QA engineers can go one step further to ensure tests. Automated testing enables your team to continuously integrate changes and rapidly execute quality checks against those changes. Start automating by looking for the areas that are:

- Repetitive
- Labor intensive
- Prone to defects

If you're starting from scratch and don't currently have any test suite, you're not alone. You have nothing to be ashamed of, but it's time to evolve and begin adopting DevOps practices that are proven to accelerate your delivery.

Your mission to build an automated test suite should start with prioritizing the areas of your codebase that have the biggest impact to customers. Which are the features or areas of logic that are most often hit while the average user is interacting with your product?

Treat the issue of building out a robust test suite as you would any other type of technical debt that you have to slowly pay back. Create tasks specific to implementing an automated test framework and write tests to provide coverage for the areas of your codebase that are the most vulnerable to breaking. Schedule time in your Agile sprints or project workflow to ensure that the work is prioritized, and then slowly add it in.

Building the tooling required for testing as well as developing the habit of writing tests for new features takes time. These are not overnight tasks, so prioritize and slowly work through it.

Testing in Different Environments

The concept of quality control in DevOps applies to more than just the code. It exercises your deployment processes and architecture as well. Each target environment will have small differences that may impact how your application runs. You want to strive to make your testing or staging environments as close to production as possible so that you can establish repeatable processes in reliable environments. Staging enables you to identify and resolve any issues with the process or infrastructure, making it easier to identify and fix changes that break any part along the way.

No ubiquitous standard exists for naming environments. Nor is a set number of environments used by every team. Every deployment process is unique to the organization implementing it.

If you're diligent about tooling and resource parity, you can force issues to surface early in the development life cycle through tests. If you're not diligent in these areas, you'll pay the price by having more issues to deal with after you release code to production (not to mention the frustration created when developers repeatedly have tickets returned to them).

The environments and steps your code travels through on its way from development to production is called the *release pipeline.* Although the release pipeline can vary because of many factors, including your application, organization, and existing tool set, a typical architecture consists of five environments:

- Local
- Development
- Testing
- Staging
- Production

Each of the four environments preceding production serves to challenge the code against increasingly difficult (and expensive) tests to ensure that the code is production ready:

- **Local:** Does the feature work in isolation?
- **Development:** Does the feature play well with the other components in the service? Does the feature respond as expected when connecting with external services?
- **Testing:** Is the feature free of security concerns? Does the user experience meet feature requirements and development standards?
- **Staging:** Does the feature meet or exceed all business requirements?

TIP

Some teams add a sandbox environment to test experimental ideas. Also, many developers work on a local environment that's unique to their machine. Keep reading for more about these various environments.

Local environment

A local environment is a single developer's machine (laptop or desktop). One of the advantages of developing and running code locally is that you don't need the Internet to run your software. The phrase "Works on my machine!" is spoken by a developer who has functionality on their computer even though the code may break in another environment. This discrepancy can happen because environments can have vast differences in technical dependencies, data, and other resources.

Require developers to write unit tests to accompany each component they write. Depending on the feature and how much it interacts with other components (in your system or third-party services), integration tests with stubbed responses may also be written and run locally.

Sometimes you need to interact with other services and tools through HTTP requests when working on your local machine. If you need to work offline, those responses can usually be stubbed. In other words, you can trick your algorithm into thinking that it received a response. Stubbing or mocking is especially important to use in your automated unit tests (refer to "Going beyond the Unit Test," later in this chapter) to speed the time that tests take to run and to ensure a consistent response for the code to ingest. Remember to update your stubbing if the API you're calling changes its response!

Development environment

The development environment is where the first phase of testing for new code takes place. This environment is often referred to as "DEV." After developers know that a feature works on their local machine, they deploy new code to DEV to test it there.

When the code is in DEV, engineers run unit tests and integration tests to ensure that the new code still works as expected when merged into the main master or trunk branch in git. Developers often also play around manually with the new functionality to double-check that it's ready for a code review by a peer and deployment to the testing environment. In other words, the development environment is where developers can determine whether they think they've accomplished what they needed to do or they need to rework it.

The development, or DEV, environment is the least stable environment in the release pipeline. Changes are constantly being integrated by developers working on multiple areas of the codebase. Developers must confirm that the code works and the tests pass consistently before passing it on to the next environment.

Testing environment

This stage is sometimes referred to as quality assurance (QA). Traditionally, after a developer felt confident in their work, they would submit a pull request to check in their code, undergo a code review, and then hand the code over to the QA team to test it in the testing environment. But that's not very DevOps-like. In DevOps, people work together and share responsibility.

Depending on how far along in your DevOps transformation you are, the QA team may still "own" the testing environment. Although this situation isn't ideal, it's a fine place to start. QA teams commonly fear automating themselves out of a job. Reframe the opportunity to show the QA team how they'll transition from the reactive and rote toil of manual testing to becoming experts in automated testing and continuous integration.

DevOps fundamentally changes the role of QA on an engineering team. No longer does a QA engineer "own" the testing environment, test code, and then pass it off to operations after it's deemed functional. Instead, DevOps empowers people in QA to act more like engineers. Today, QA teams assist in writing automated tests and serve as experts in testing practices, procedures, and approaches.

The DevOps emphasis on automation and continuous improvement make the hand-off to QA more nuanced. As you consider how DevOps will impact your testing practices, take time to think through what your QA team might look like in the next year. How will you level up your QA engineers? And how will you take advantage of their unique knowledge to teach developers how to write better, more reliable tests?

No matter who deploys the code to the testing environment — or whether deployment happens automatically in a CI/CD setup — it's a slightly more robust environment than development (more resources and data) in which additional tests are run. Although unit tests can verify functionality in logic, they lack the whole picture. The testing environment is an ideal place to start running user interface tests and security challenges.

REMEMBER

Tests may be run in two ways: serially, with each test being run sequentially, one at a time; or in parallel. A parallelized testing environment is advanced but is a differentiator between high-velocity engineering teams and those with slower software delivery.

Staging environment

The staging environment should be a mirror of the production environment. These two environments should have data and resource parity (or as close as you can get) so that you can confirm that the infrastructure does not have an unexpected impact on the code being released. The only difference between staging and production is that staging does not serve customer traffic. This approach enables you to ensure that the code is performant, and you can check for potential bugs with external services and database interactions. In addition to being the place for final testing, staging is where certain configuration or migration scripts can be run.

TIP

Although staging should mirror production as much as possible, it will never fully emulate the production environment because it lacks customer interaction and usage. Different approaches to testing in production and releasing software (discussed in the next chapter) have evolved from this fact. Testing isn't fail-safe, but it will give you and your team confidence in your software and limit the blast radius of potential failures.

Production environment

The production environment is the final stage for your code, and it's the one in which you have the most to lose. Your production environment serves customer traffic. After a build is released to production, it's supposed to work as expected. Of course, in the real world, things go wrong all the time. As long as you have a way of handling rollbacks or deploying in a phased manner, you should be fine. (I discuss deployment approaches in Chapter 11.)

Being notified by customers of an incident is not ideal because it damages trust. Application insights, monitoring, logging, and telemetry are all tools that provide you with information on your system's on performance, server load, and memory consumption. Ideally, your incident alerting system (discussed in Chapter 17) brings issues to your attention before your customers reach out. Even so, make sure that your customers can easily get your attention when they're impacted.

Going beyond the Unit Test

In unit testing, developers make sure that each component does its job and then continues to do its job after updates and changes. But what happens when those components get combined? And what happens when they are migrated to the next environment in the pipeline?

Your development life cycle should include time for the following:

- Developing test cases
- Writing automated tests
- Running manual tests (if still required)
- Reflecting on the delivery
- Making adjustments

I highlight some of the most insightful and critical tests to include in your automated test suite in the following sections. It's far from an exhaustive list, but it will get you started on your path to continuous testing and serve as a baseline as you continue to grow and refine your approach to testing.

Unit tests: It's alive!

Developers write unit tests as they work to test the functionality of the logic they just built. A single function may have a dozen associated tests. Just as functions should do only one thing, so, too, should tests. Each test should ensure that the algorithm works as expected through a variety of scenarios.

Unit tests give developers immediate feedback and eliminate multiple loops of the traditional development life cycle. Instead of writing code, passing it to the QA team, and having them kick it back repeatedly, an engineer can check their work within seconds.

Unit tests are cheap, meaning that they require fewer dependencies (they test the functionality of only one piece of code) and they run quickly. A unit test can run in milliseconds, as compared to certain user interface or end-to-end tests that, depending on the complexity of the component, can take minutes to run.

WARNING

Code coverage refers to how much of your codebase is "covered" by tests. Many tools evaluate your codebase against your test suite and give you a percentage of coverage, but that approach is flawed because it doesn't measure the quality of those tests and is easily gamified. I think code coverage is more useful as a data point for stubborn executives than as a real measure of the efficacy of your engineering team. Trust them to write quality tests and to verify that work in code reviews. Provide continuing education opportunities for engineers to share knowledge and learn how to write better tests, not just more tests.

Integration tests: Do all the pieces work together?

Integration tests are typically the most useful in staging (see the "Staging environment" section, earlier in this chapter), where the application has access to the network, databases, and file systems. Unlike unit tests that validate functionality of a single piece of logic, integration tests confirm that multiple components communicate as expected.

Though a bit more complex than other tests to set up, integration tests catch bugs that are hard to track down. Not only do all the pieces of code need to work together, but they have to work with the rest of the environment as well. In integration testing, you are looking for all the little variables that can make things go awry. How does the code work with real data? What about with heavy user traffic? Do problems arise when the code interacts with mail servers?

Stubs are snippets of code that mimic a user action in a test. Drivers, on the other hand, mimic a server response.

Regression tests: After changes, does the code behave the same?

Regression testing verifies that after you make changes to the code, key metrics for how your application works and runs haven't changed as well. This verification includes previous functionality. Have old bugs resurfaced? Did a new change impact a previous version of an API?

This testing might check that the accuracy or precision hasn't degraded. Sometimes regression tests are as simple as ensuring that a simple CSS color change didn't make the site a different color or cause a link to break.

Visual tests: Does everything look the same?

Visual testing is relatively new and fascinating. It's essentially automated testing for the user interface (UI) and ensures that the application appears the same to users (tailored to specific browsers and devices) — down to the pixel. Every other kind of test verifies an expected function. Visual tests are unique in that they test the UI for consistency. I highly recommend that you don't roll your own visual testing tool and instead opt for one of the dozens of open source or enterprise tools available.

Visual testing works by establishing a visual baseline through a screenshot, which serves as the expected display. When you merge a change into the master code branch, the testing library will take a screenshot of the new results and compare it to the baseline. If the test detects differences, the test fails. Some tools even go so far as to highlight the differences so that you can see exactly what changed — which is a front-end developer's dream.

Performance testing

Performance tests verify the overall application performance. Is the app responsive? Stable? Does it scale as expected and use a reasonable amount of resources? Performance testing can also include security tests and load tests. Security tests verify that no known vulnerabilities were introduced in the latest build, and load tests mimic a large number of users or data that will stress the system.

TIP

Don't forget security! Security tests should look at network security and system security as well as client-side and server-side application security. The world of security testing is vast and deep. I highly recommend the Open Web Application Security Project (OWASP) testing guide found at `https://www.owasp.org/index.php/Category:OWASP_Testing_Project`.

Continuous Testing

From a developer perspective, testing has traditionally been overlooked. DevOps, however, emphasizes the importance of testing. As developers deliver software faster and in an automated fashion, the quality of the work can't degrade. Mistakes can be costly.

An untested and buggy release can have a permanent impact on your reputation or open you up to security and compliance risks. Although continuous delivery and continuous integration are more well known than continuous testing in DevOps, continuous testing is finding its place.

Continuous testing starts in the development stage, and developers can spearhead its use in order to get immediate feedback on their work and prevent late nights resulting from incidents and outages. When organizations embrace DevOps, taking care of quality becomes everyone's job — not just QA's.

Continuous testing can guide software development teams when it comes to meeting their business goals, managing business expectations, and providing data for decisions that require a trade-off. As with many things in DevOps, continuous testing will shorten your cycles and enable you to rapidly iterate.

No matter what approach you take to testing, your code will need to make its way to production eventually, and how you deploy a product is the subject of the next chapter.

» Implementing CI/CD

» Managing deployment incidents

» Choosing a deployment style

Chapter 11
Deploying a Product

As I discuss in Chapter 10, you should thoroughly test all code before releasing it to customers. The deployment process refers to releasing that code to customers. That process can be as simple as clicking a button or as complex as a series of pipelines and gates through which the code must pass to reach customers. Sometimes you hear the term *release* used interchangeably with *deployment.*

A *deployment* is the movement of code from one environment to another. A developer can deploy their code from their local machine to the development (DEV) environment. At that point, the code may pass through several more environments, like user acceptance testing (UAT) or quality assurance (QA), staging, and production (PROD). The deployment to production — specifically, a deployment to customers — is the purest form of the word *release.*

In this chapter, you find out how to implement continuous integration and continuous delivery (CI/CD), decide on a deployment strategy, and manage releases.

Releasing Code

If code is accessible by customers, it has been released. If code is exposed to a new environment, it has been deployed. Still, deployments have enough shades of grey to render this bifurcation of meaning unhelpful. So although the terms *released* and *deployed* aren't perfectly synonymous, for the purposes of this chapter, I use them to mean the same thing.

Releasing a build (an artifact of the packaged code) to the production environment does not necessarily mean that is serving all customers — or any customers, for that matter. *Release* simply means that a version of the application is now receiving production traffic and has access to production data.

I've seen more than one conversation get sidelined over the use of *release* and *deployment* to mean different things. As with all things in DevOps, communication is key. Never assume what someone means by either term without asking for clarification.

Another term often thrown around is *shipped*. This term derives from the time when companies literally shipped CDs to customers with updated software for installation. In fact, developers will often joke with each other when one asks about whether something is ready to be released to customers. "Ship it!" they say.

Although the origins of "shipping" software typically referred to delivering a new version of software to customers, people use it interchangeably with *deploying* and *releasing*. The bottom line is that the meaning of all these words depends on the message intended by the person using them. If you're unsure, ask.

Integrating and Delivering Continuously

The growth of DevOps culture has changed the way developers build and ship software. Before the Agile mindset emerged, development teams were assigned a feature, built it, and then forgot about it. They tossed the code over to the QA team, who then threw it back because of bugs or moved it along to the operations team. Operations was responsible for deploying and maintaining the code in production.

This process was clumsy, to say the least, and it caused quite a bit of conflict. Because teams existed in silos, they had little to no insight into how other teams operated, including their processes and motivations.

CI/CD, which stands for continuous integration and continuous delivery (or deployment), aims to break down the walls that have historically existed between teams and instead institute a smoother development process.

Benefitting from CI/CD

CI/CD offers many benefits. However, the process of building a CI/CD pipeline can be time consuming, plus it requires buy-in from the team and executive leadership.

Some benefits of CI/CD include:

- **Thorough automated testing:** Even the most simple implementation of CI/CD requires a robust test suite that can be run against the code every time a developer commits their changes to the main branch.
- **Accelerated feedback loop:** Developers receive immediate feedback with CI/CD. Automated tests and event integrations will fail before new code is merged. This means that developers can shorten the development cycle and deploy features faster.
- **Decreased interpersonal conflict:** Automating processes and reducing friction between teams encourages a more collaborative work environment in which developers do what they do best: engineer solutions.
- **Reliable deploy process:** Anyone who's rolled back a deploy on a Friday afternoon can tell you how important it is that deploys go smoothly. Continuous integration ensures that code is well tested and performs reliably in a production-like environment before it ever reaches an end user.

Implementing CI/CD

CI/CD is rooted in agile methodologies. You should think of implementing CI/CD as an iterative process. Every team can benefit from a version of CI/CD, but customizing the overall philosophy will depend heavily on your current tech stack (the languages, frameworks, tools, and technology you use) and culture.

Continuous integration

Teams that practice continuous integration (CI) merge code changes back into the master or development branch as often as possible. CI typically utilizes an integration tool to validate the build and run automated tests against the new code.

The process of CI allows developers on a team to work on the same area of the codebase while keeping changes minimal and avoiding massive merge conflicts.

To implement continuous integration:

- **Write automated tests for every feature.** This prevents bugs from being deployed into the production environment.
- **Set up a CI server.** The server monitors the main repository for changes and triggers the automated tests when new commits are pushed. Your CI server should be able to run tests quickly.
- **Update developer habits.** Developers need to merge changes back into the main codebase frequently. At a minimum, this merge should happen once a day.

Continuous delivery

Continuous delivery is a step up from CI in that developers treat every change to the code as deliverable. However, in contrast to continuous deployment, a release must be triggered by a human, and the change may not be immediately delivered to an end user.

Instead, deployments are automated and developers can merge and deploy their code with a single button. By making small, frequently delivered iterations, the team ensures that they can easily troubleshoot changes.

After the code passes the automated tests and is built, the team can deploy the code to whatever environment they specify, such as QA or staging. Often, a peer manually reviews code before an engineer merges it into a production release branch.

To implement continuous delivery:

- **Have a strong foundation in CI.** The automated test suite should grow in correlation with feature development, and you should add tests every time a bug is reported.
- **Automate releases.** A human still initiates deployments, but the release should be a one-step process — a simple click of a button.
- **Consider feature flags.** Feature flags hide incomplete features from specific users, ensuring that your peers and customers see only the functionality you desire. (I discuss feature flags more later in this chapter.)

Continuous deployment

Continuous deployment takes continuous delivery even one step further than continuous delivery. Every change that passes the entire production release pipeline is deployed. That's right: *The code is put directly into production.*

Continuous deployment eliminates human intervention from the deployment process and requires a thoroughly automated test suite.

To implement continuous deployment:

- **Maintain a strong testing culture.** You should consider testing to be a core part of the development process.
- **Document new features.** Automated releases should not outpace API documentation.
- **Coordinate with other departments.** Involve departments like marketing and customer success to ensure a smooth rollout process.

Managing Deployments

Release management is a core component of DevOps and an area in which you're likely to see the most improvement when adopting DevOps practices. As mentioned elsewhere in the book, developers and operations folks used to be isolated from each other, existing in silos of knowledge and responsibility. Developers wrote code, added functionality, and then tossed it to operations for deployment and maintenance — all without properly communicating technical considerations important to the release.

Often, manual deployments, compounded by poor collaboration, lead to less-than-stellar outcomes. In 2016, the research company Gartner estimated that a lack of effective release management contributed 80 percent of service outages in large organizations.

Releasing software in an automated and well-orchestrated fashion is key to reducing service outages and incidents.

Automating the right way

Although automation is key to accelerating your software delivery, use caution when automating your release processes. You need ensure that you're automating the appropriate procedures. The worst thing you could do is to abstract a problematic process and implement it in a way that removes humans from the process. High-performing engineering organizations use automated tooling in their release processes, but they take a lean approach, adapting the tooling as necessary.

If you have a relatively small engineering organization, I recommend standardizing release and deployment processes across the company, at least at the start. Your release procedures will change and evolve as you grow. Organizations like Amazon assign a site reliability engineer or operations specialist to each engineering feature team. Because so much of Amazon's infrastructure and architecture is microservice-based, those teams can operate independently. Until you feel that your team is at this level of performance, keep your release and deployment processes consistent.

Versioning

You version software upgrades by assigning a unique version name or version number to identify different states of an application. You can even differentiate states of source code internally looking at the code commit history in git. You can identify and even select the previous states of the code — the revision history using the unique SHA-1 hash that accompanies every commit.

Versioning deployments is equally important. If you utilize CI/CD, you should check version numbers identifying software state into your source control.

Semantic versioning

Ad hoc versioning never goes well. The various humans on your team all think differently from each other, and those subtle differences — without versioning standards — can lead to confusion. Semantic versioning is a relatively simple approach that everyone on your team should be comfortable getting behind.

The real benefit of semantic versioning is how the version number gives you important information when viewed in relation to the version numbers of the preceding and subsequent releases. The actual version number distinguishes patches from minor releases and major version updates by how the version numbers increment.

Semantic versioning uses three numbers in every version number. The number that increments depends on the type of release. For example, the currently released version of ACME APP is 1.3.4, so here's what various versioning would look like:

- A patch update would make the current release 1.3.5.
- A minor update would increment to 1.4.0.
- A major update would put the release version at 2.0.0.

The term *patch* refers to a deployment that fixes bugs. The changes are minor and simply reinforce previously released functionality. Minor version updates contain new features. Major updates aren't backward compatible and include code that would break previous versions.

This system helps you easily track versions internally as well as inform your users, depending on how and when you choose to announce releases publicly.

Versioning for continuous deployment

Semantic versioning gets a little more tricky than how I describe it in the preceding section if you're deploying ten times per day — or even once per day. It's complicated when you quickly increment while having extremely minor differences between versions released.

At this point, I recommend adding a dynamic component to your versioning. Because continuous delivery and continuous deployment are automated, a code check-in will trigger a new build. When completed, that build will then trigger a release pipeline that deploys the build to the various environments. Every releasable build should have a unique version number.

Variables enable you to implement more complex versioning while still maintaining a semantic approach. Build tools allow for you to add global or build variables to a release number, thus distinguishing it from the others.

Most automation tools permit variable groups that set the values and definitions across the entire release pipeline. You typically format variables like this: ${variable}. The pipeline tooling helps you ensure that no two releases are identically named. Here are some examples:

- ${developer}: v1.3.4-efreeman
- ${team-project}: v1.3.4-serverless
- ${email}: v1.3.4-emily@microsoft.com
- ${commit}: v1.3.4-bc0044458ba1d9298cdc649cb5dcf013180706f7

Depending on the tool you're using, you can get extremely granular and mix and match variables however you like. I advise adding enough information to uniquely identify state and provide context to reviewers while maintaining readability:

- v1.3.4-serverless-emily@microsoft.com
- v1.3.4-release-54-bc0044458ba1d9298cdc649cb5dcf013180706f7
- v1.3.4-efreeman-critical-security-patch

Tracking application packaging

Releasing microservice architecture and distributed systems involves significantly more moving parts than deploying a monolith. As a result, you can't simply track the state of each service or component; you must track the entire application as a package, including all the components and database changes.

If you have different components of an application deployed to various containers or clusters, deploying each piece every time a new version is released is wasteful, and risks the possibility of errors. Instead, you need to use a configuration management tool to track the deltas — that is, the changes and differences between versions. If an element of a component changed, you rerelease the component. If not, you leave it in its current (and up-to-date) state. This approach minimizes downtime and reduces failure.

TIP

Standardizing infrastructure configuration allows developers to stand up new infrastructure (servers, containers, VMs) without the assistance or approval of an operations specialist, empowering developers with more autonomy and allowing them to take more ownership of their work.

Mitigating Failure

No other activity opens a development team to failure as much as deployments and releases. That risk of failure is one of the reasons that traditional engineering teams avoided deployments and made them occur as infrequently as possible. Releasing software was a headache — one that they wanted to avoid.

But that avoidance of deployments is what causes a lot of the problems that occur with them. You improve on the activities that you do frequently. Frequent deployments mean smaller changes. A few dozen lines of code are less likely to cause service interruptions than heavy amounts of code. Finding bugs in small releases is easier than digging through hundreds of lines of code in dozens of files.

No matter how frequently you deploy or how you approach deployments and releases, they can cause failure. You can use DevOps to mitigate that failure.

Rolling back

Rolling back is the easiest and most frequent way of restoring service after a deployment outage. You essentially roll the current deployment version back to the last stable build. You have two ways to do this: restore a previous deployment or create a new deployment with a unique identifier as a copy of the previous stable version.

A rollback is called for when a build is released and breaks the production environment, likely impacting customers. If application performance or availability is impacted, the quickest fix could be to redeploy a previous version known to be stable. Other times, teams choose to troubleshoot live and create a hotfix in real time. That approach isn't typically ideal for customers or engineers because it's stressful, to say the least. However, as the next section explains, sometimes it's the only viable option.

Cloud providers can enable you to quickly roll back using release pipeline tools.

Rollbacks are typically initiated manually. Automation tooling can use monitoring thresholds on performance and other application metrics to detect a potential problem and alert engineers. If you're using a release pipeline, a rollback is sometimes as simple as a click of a button.

Fixing forward

Occasionally, rolling back isn't possible. Most often, database changes make it difficult or impossible to simply move backward in builds. If you release a new schema, migrate data, and allow customer data to populate the new columns, you've got a challenge ahead of you.

In the scramble to fix a production issue on the fly, you risk breaking other functionality, accumulating technical debt, and hindering development of other engineers by freezing work while fixing forward. I recommend taking this approach if it's the only option you have. Then, use your post-incident review to explore architecture changes that would ease recovery for future outages.

Democratizing Deployments

Traditional engineering organizations commonly had deployment roles — even release engineers who specialized in deploying software. That is not an ideal approach because it strips power from the team as a whole and silos responsibility. Remember, in DevOps, you share as much information as possible. Specializing in specific areas of engineering or having an expert in a particular language, framework, or tool are absolutely reasonable, but you want to avoid making the "expert" the only individual capable — or allowed — to do a specific job.

Enough tools are available today for literally anyone to be capable of learning how to package and release your application. If your release process is so complicated that only two people can manage it by following 13 pages of instructions, then it's time to start at the beginning and completely redo your release process.

REMEMBER

I talk about transitioning to the cloud in Chapter 21. That move, although potentially time-consuming, is an excellent opportunity to revamp and modernize old processes. Just because you've always done it a certain way doesn't mean that it's going to carry you into the next phase of your business. Operations is accelerating, and you must adapt to remain competitive.

It's okay if you have specific security or compliance concerns that don't allow you to move to continuous deployment tomorrow. Please don't get overwhelmed with what you're "supposed" to do. Instead, evaluate where your organization is realistically and then make a plan to continuously improve and adapt.

Many companies simply aren't capable continuous deployment, nor are they willing to allow new code into production as soon as it has been merged. It requires an enormous amount of upfront work to build robust testing, security gates, and pipelines. I don't want to understate that. This isn't easy.

REMEMBER

CI/CD is the end goal, but the journey is equally, if not more, important. Slowly move your team toward continuous integration and continuous delivery — and forget the pressure to modernize overnight. Remember, if you attempt to change everything too quickly, your DevOps transformation will fail. Accept where you are and make a plan to grow and continuously improve.

As you adopt CI/CD, it's absolutely fine to create human gates in release pipelines to ensure quality — especially as you're getting used to this new approach. Just be sure to apply reliability calculations to your people as well as your machines. Select three people ($n + 1$) who can approve builds for release. This approach allows one person to go on vacation and another to get sick without creating a bottleneck in engineering productivity. (See Chapter 3 for details on dealing with bottlenecks.) You want to remove bottlenecks, not create them.

Choosing a Deployment Style

Many approaches to releasing software to customers are available, and the practices considered to be good have evolved. Choosing a deployment style is where infrastructure knowledge becomes much more important for your engineering team. It's also why I'm vehemently against NoOps.

TIP

NoOps — short for no operations — is the suggestion that automation can and should replace operations specialists. This idea is foolish because no matter how robust your automation becomes or how much you abstract the underlying infrastructure for developers, core infrastructure and operations knowledge will always be vital to an engineering team.

The operations people on your team are experts in infrastructure. They understand the history of system administration and hosting software — which provides context for the deployment styles we think of as ideal today. Software infrastructure has built upon itself and adapted to new challenges.

Deployment styles are no different. You have plenty to choose from, and each one has advantages and disadvantages. But the options described in the following sections are intended to minimize the risk of negative customer impact.

Blue-green: Not just for lakes

Blue-green deployments are one of many release styles that seek to reduce service outages resulting from a bad deployment. In this case, blue and green have no particular meaning. They could just as well have been called pink-red deployments or yellow-purple deployments. This name is simply a way of identifying the two versions of your application running in production.

And that's just what blue-green deployments do — release two versions of your software to the production environment. You utilize a router to determine which version customers have access to.

Imagine that the current release running in production is v2.0.4. Everything's going great and you're ready to release a minor update, which will take you to v2.1.0. Before you release the new version, only v2.0.4 is running in production, as shown in Figure 11-1.

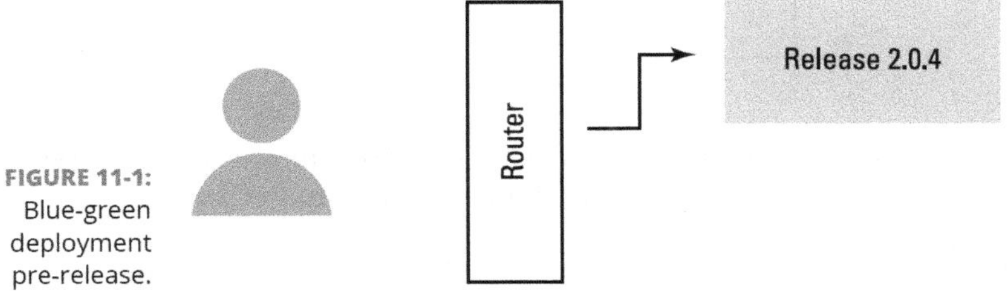

FIGURE 11-1: Blue-green deployment pre-release.

To ensure that the new version behaves as expected in your production environment — without negatively impacting customers if something goes poorly — you release v2.1.0 to production but route all traffic to the stable v2.0.4. You can see what this looks like in Figure 11-2.

Both versions are running in production, but nothing has changed for customers. You can leave the new release running in production for as long as you like (taking into account resource consumption). I recommend running tests on v2.1.0 in production and ensuring that everything performs as expected.

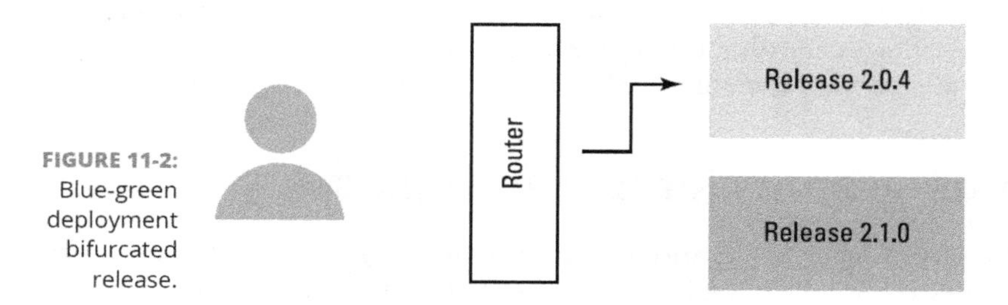

FIGURE 11-2: Blue-green deployment bifurcated release.

After you're confident that the new version of your software is stable and ready for customer traffic, it's time to make the switch. The router will then trigger all customers to reach the latest stable version (v2.1.0) and stop sending traffic to the previous release (v2.0.4). At this point, your production environment will look something like Figure 11-3.

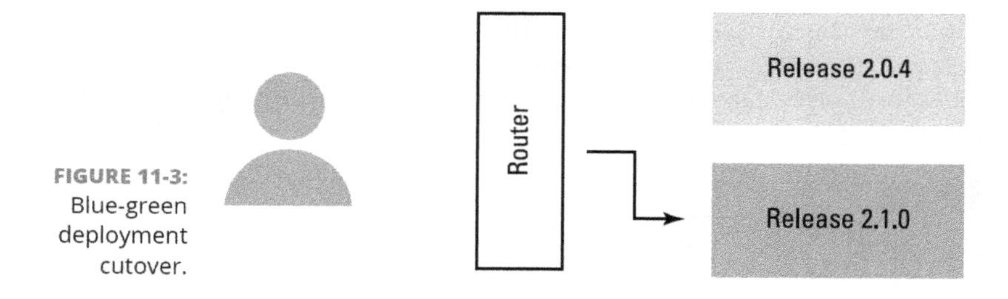

FIGURE 11-3: Blue-green deployment cutover.

You can feel confident that the new release is stable and won't cause service outages or performance impact when customer traffic is routed to it.

For me, one of the most important benefits of a blue-green deployment is the ease of rolling back. The latest stable version is already running in production. You simply need to reverse the router cutover and send traffic back to the previous version.

Schrodinger's canary: The deploy's not dead (or is it?)

Before modern detection tools came into use, coal miners took canaries into the mines with them. If poisonous gas began to collect, the canary was the first to die.

Its death warned the humans to the danger and initiated an evacuation of the area. Macabre, to be sure, but effective.

Luckily, no canaries are harmed in the process of a canary deployment. This style of release takes blue-green deploys one step further: It slowly transitions between the two versions rather than cutting over all at one time. Canary releases ship software changes to select customers as a way of testing functionality and reliability in production while limiting the number of customers potentially impacted.

Refer back to Figure 11-2, which shows the blue-green deployment bifurcated release. You've released the newer version, v2.1.0, into a production environment but the router is blocking traffic. If everything looks good after a period of time, you're ready to begin slowly introducing customer traffic.

Unlike a blue-green deploy, the router will send customer traffic to *both* versions until 100 percent of traffic is directed to the new version. The number of customers (or type of customer!) you select to be the canaries is up to you. I recommend starting with a percentage, but some companies prefer to select customers based on demographic information or location. The latter is useful if you're deploying a new version of your application to a specific region first.

Imagine that you decide on 10 percent. You direct the router to send 10 percent of customer traffic to the new version, as in Figure 11-4, which shows the start of a canary deploy. After you're satisfied that no negative customer impact is occurring, you can slowly increase the number of customers who receive the new version of your application. How quickly you deploy the updated version or how many customers are included in each chunk is completely up to you. It can be as smooth and slow as you like.

TECHNICAL STUFF

You can tailor canary deploys to the type of release. If it's a bug patch, you'll likely be able to release much faster, whereas with a major update, you'd be wise to take your time when increasing traffic.

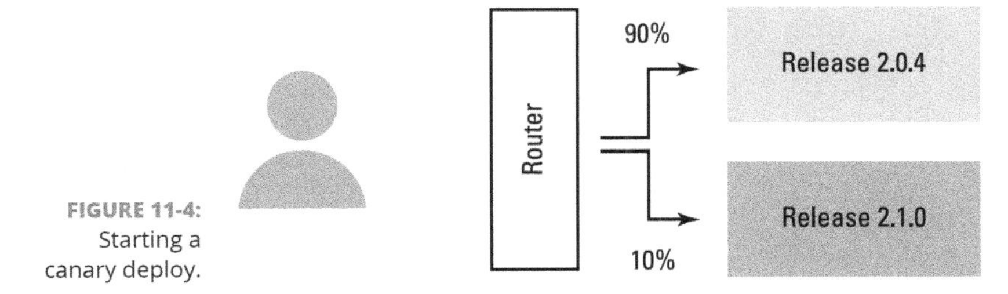

FIGURE 11-4: Starting a canary deploy.

Many companies choose to dogfood their own products — meaning that they use the software they sell internally. Canary deploys (and the related option of feature flagging I discuss in "Toggling with feature flags," later in this chapter) offer a uniquely wonderful solution. You can deploy the new version to selected users and test the functionality included in the update for days or weeks before releasing to your entire customer base.

Rolling the dice

The final type of release I want to highlight is called a rolling deployment. Instead of releasing a new version to select customers in small incremental chunks, rolling deployments replace the version of an application running on a specific instance. The new version is deployed to each instance one at a time (or in clusters) until all instances or machines are running the latest version.

Some companies choose to implement rolling deployments by cutting over multiple machines at the same time. The size of your grouping is referred to as the *window size.* A window size of one will proceed one machine at a time whereas a window size of four will deploy the new version to four servers at the same time. Figure 11-5 shows what the beginning of a rolling deployment might look like.

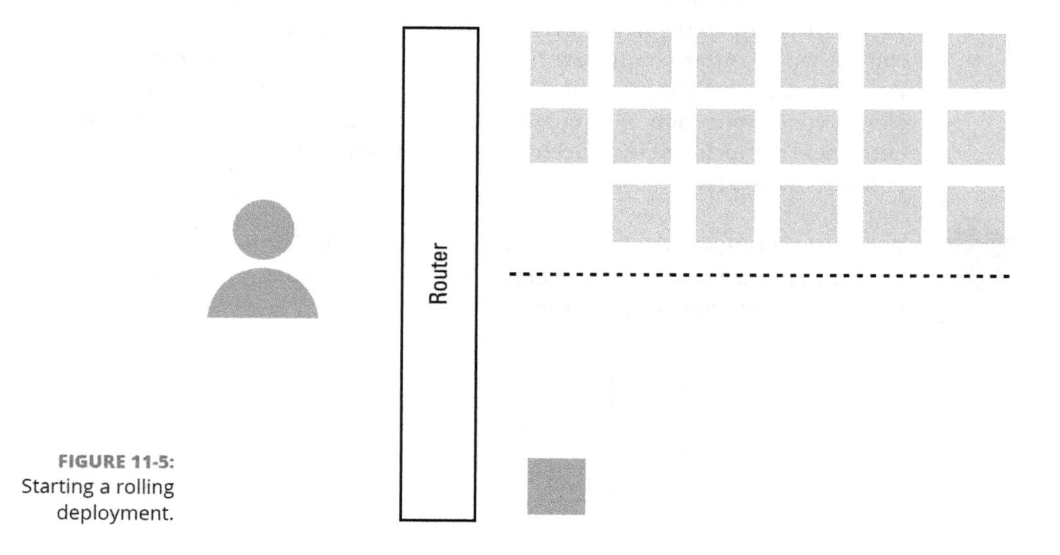

FIGURE 11-5: Starting a rolling deployment.

The real advantage of a rolling deployment is the contrast between it and a traditional upgrade. Historically, you would have to take all servers offline and deploy the update — praying everything went well.

Figure 11-6 gives you an idea of what your system will look like toward the end of a rolling deployment.

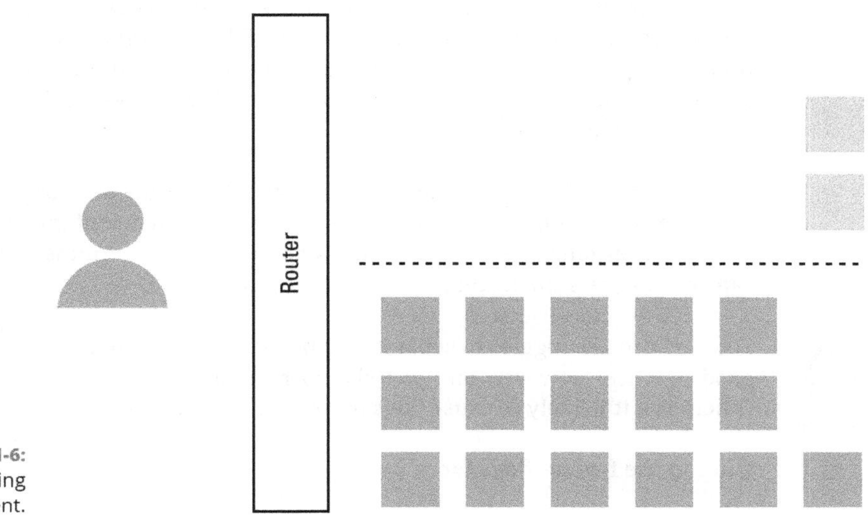

FIGURE 11-6: Finishing a rolling deployment.

Now, with a rolling deployment, you can use a load balancer or router to direct traffic to the servers still running the current version of your application. The machines being upgraded won't see any traffic until you're satisfied, allowing all nodes to be updated with zero customer impact. In addition, rolling deployments require fewer structural resources than blue-green deployments and canary releases.

TIP

I recommend that you take time to think about how sessions will persist during a rolling deployment.

Toggling with feature flags

A feature flag or feature toggle is a conditional feature that can be hidden from customers. This is an excellent solution for maintaining continuous delivery or continuous deployment while not releasing the functionality to customers until you're ready.

TECHNICAL STUFF

Feature flags enable marketing and sales organizations to release a feature or feature set to customers with a coordinated release while not impacting engineering's ability to continuously develop and deploy the functionality.

If you and your team feel extra radical, you can deploy partially completed work to production. Deploying this way reduces the number of feature branches and merging you have to manage throughout the process and alerts other developers of your work long before it's finished. With a toggle, you're in complete control of user access. Although the feature code is visible to developers, the actual functionality is hidden in the user interface until you decide to reveal it. You can even select to reveal it to certain users — like internal testers — while keeping it hidden from customers.

If using feature flags, a developer first assigns the feature as a toggle either in the database (0 for "off" and 1 for "on" works well) or in a configuration file. Then the developer builds out a conditional statement that determines whether a user can see or access the feature from the user interface.

WARNING

You can also use toggles to separate old business logic from new code. This isn't a good practice, however, and you should refactor or delete old code if possible. Otherwise, it's likely to cause bugs or undesired outcomes in the future.

You can use feature flags for

- Releasing new features in development to specific users
- Updating or enhancing current functionality
- Disabling or deprecating a feature
- Extending an interface

Monitoring Your Systems

After you release the software, you need to monitor your software for performance, availability, security, and more.

Understanding telemetry

Telemetry is just a fancy word for collecting data on the behavior of your systems. Telemetry enables your system to regularly update you on how things are going, which keeps you from digging into logs only when something goes wrong. Telemetry creates records on its own behavior independently.

The real benefit here is you have a working baseline for application and system performance. If you release a new version of your software, you can watch your telemetry and look for odd behavior. If you see a spike in load time seconds after the release, it's a good indicator that something went wrong.

Telemetry is also handy in the case of a service-level agreement (SLA), which is essentially your promise of availability to customers. An SLA is typically a legal contract that promises a certain level of performance, such as 99.9 percent availability. Telemetry can help you track whether you're meeting your objectives and communicate appropriately with customers.

Recording behavior

To benefit from telemetry, you must set your application and infrastructure up in such a way that data collection and reporting are possible. Telemetry requires two components:

- **Data collection:** High-performing DevOps organizations collect data on hundreds, if not thousands, of key indicators. The metrics originate at every layer of your system: the application, environment, and network operations.
- **Metrics management:** You need a central place to store and analyze the data you collect. This platform should go beyond listing events. Ideally, you'll have a way of visualizing the data and highlighting trends. You can integrate this capability into your alerting system to ensure that engineers are notified of potential problems.

If you're not sure what to store, I suggest starting with customer events and system performance. Examples of customer events are the number of logins, sales numbers, and page load times. If sales suddenly stall during the busiest time of the day, you want to know as soon as possible. System performance includes database performance, network operations, CPU, and security.

After you have a gauge for telemetry and how it fits into your system, you can expand the areas in which you collect data:

- Number of new user signups
- Completed sales
- Abandoned checkouts
- Monthly recurring revenue (MRR)

- Response times
- Number of exceptions
- Server traffic
- Disk usage
- Deployment lead times
- Deployment frequency

Telemetry provides you with insight into your entire system at every layer and in each component. This insight ensures that you catch small issues before they cascade into system errors and reduces the frequency of having customers alert you to service outages.

I recommend that you categorize telemetry data and make it easy for your engineers to dig into various areas of your system. You can even do so by categorizing data by urgency: DEBUG, INFO, WARN, ERROR, and FATAL.

SLAs, SLIs, and SLOs

Site reliability engineering, which is a prescriptive and operations-focused interpretation of DevOps, uses three terms that are important for you to keep in mind when monitoring your systems. Each measurement assists you in determining whether your team is meeting business objectives from an engineering perspective.

- **Service-Level Agreement (SLA):** Availability — your service being up and running — is key to success in operations. Customers have to be able to access your applications. In other words, is the application functioning as expected? A service-level agreement is the availability you agree to maintain with your customers over a set period of time. It is typically part of your service contracts and is legally enforceable.

WARNING

If you set an SLA, be careful that it is not too rigid. You can easily promise a certain level of uptime and realize down the road that it's simply impossible to maintain (especially given how much your clients pay for the service). Many companies offer SLAs only to enterprise clients to ensure that they can dedicate additional resources — human and hardware — to maintaining that commitment. Violating an SLA can both cause financial consequences and damage the long-term client relationship. If you have no idea what level to start at, look at your current uptime. Industry SLAs typically range between 99.9 and 99.99 percent.

- **Service-Level Objective (SLO):** SLO is the measurement you set as what is an acceptable level of availability internally. Typically, your SLO is internal and harsher than the SLA — giving you some wiggle room. If your SLO is 99.99 percent availability, your SLA might be 99.9 percent uptime. Set reasonable SLOs. If you aim too high and overstretch your resources, you're setting yourself up for failure because you'll never meet the standard.

TECHNICAL STUFF

I refer to availability often in this section because I consider availability to be the most important measurement; other measurements, such as service latency, don't matter if the service can't even respond to a request. Keep in mind, however, that availability is just one aspect of service reliability. Refer to Chapter 8 for more information on latency, throughput, fidelity, and durability — which are all measurements of your overall site reliability.

- **Service-Level Indicator (SLI):** This is the medium through which you measure your success at meeting your SLOs. The indicators are feedback from your systems that give you reasonable insight into your actual availability percentage in each service you measure. If SLIs dip below expected thresholds (SLOs), you need to dig in further and determine whether changes to the system are necessary.

Beyond your telemetry tooling, additional monitoring tools include dashboards, logs, and third-party analytics that look for data patterns and security threads. A parsing tool for logs will allow you to more easily and rapidly gather important information as needed.

3 Connecting the Circuit

IN THIS PART . . .

Create an iterative cycle of continuous improvement and increased velocity by prioritizing critical work and improving performance.

Develop a feedback process that enables customers to quickly alert you to what they love (and hate) about a product, ensuring your ability to integrate feedback into your product road map continuously.

Hire and retain DevOps talent, and organize your engineering organization to maximize skill sets.

Permit your engineering team autonomy to scale your organization with DevOps-focused incentives.

» **Prioritizing important work over false urgency**

» **Increasing velocity through improved performance**

Chapter 12

Implementing Rapid Iteration

The term *fail fast* became a mantra of startup culture in the early 2000s and became widely used thanks to the ubiquity of Eric Ries's 2011 book, *The Lean Startup*. Facebook, one of the grand successes of Silicon Valley startups, even went so far as to make its motto "Move fast and break things" — in other words, fail fast. The fail fast mentality became popular in Silicon Valley because of its emphasis on quick innovation, something critical to companies looking to disrupt industries with novel innovations.

The original intent of the term *fail fast* was to encourage startups to build minimum viable products (MVPs) — small subsets of features designed to satisfy early adopters — to experiment, verify assumptions, and collect customer feedback before dedicating capital to large-scale projects. Innovation and iteration are tenets of DevOps, but failing too fast and too often can cause more problems than it solves.

For Facebook, this fact became so apparent that Mark Zuckerberg announced an update to the motto in 2014. Facebook now embraces, "Move fast with stable infra." (Infra refers to infrastructure.) Innovating at the cost of reliability and availability for your customers is problematic, especially if moving fast loses your company money.

In this chapter, I discuss rapid iteration, but keep in mind that "moving fast" will depend on the context and constraints in which your team operates. To understand rapid iteration, you need to prioritize important, proactive work to limit urgent, reactive responses. You need to recognize the three constraints of any project — speed, price, and quality — as well as adopt the practices of high-velocity engineering teams.

Prioritizing the Important

One of the most significant aspects of rapid iteration isn't choosing what to do next but rather choosing what *not* to do. President Dwight D. Eisenhower said, "I have two kinds of problems, the urgent and the important. The urgent are not important, and the important are never urgent." Steven Covey took Eisenhower's philosophy and created the Eisenhower Decision Matrix for his book, *The Seven Habits of Highly Effective People.* Figure 12-1 shows my version of this decision matrix.

The matrix is divided into quadrants. The upper left is important and urgent; the upper right is important but not urgent. The lower left box is urgent but not important and the lower right is neither important nor urgent.

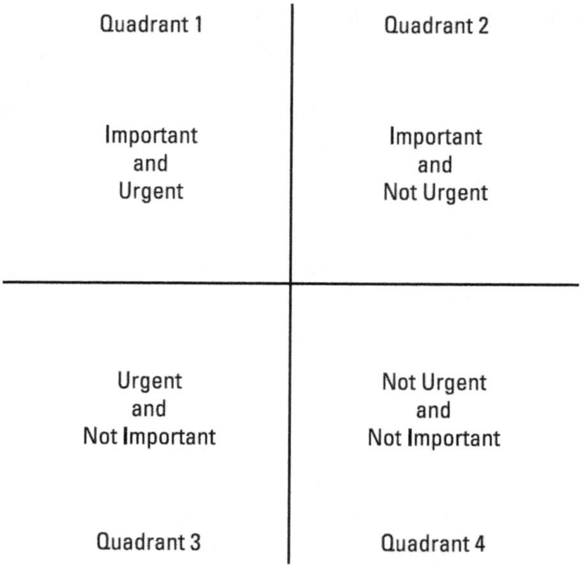

Quadrant 1	Quadrant 2
Important and Urgent	Important and Not Urgent
Urgent and Not Important	Not Urgent and Not Important
Quadrant 3	Quadrant 4

FIGURE 12-1: The Eisenhower Decision Matrix.

I love this decision matrix for engineering teams because it forces you to consider what is important to your business and what is simply noise — distractions from your mission. Dozens of distractions bombard your team every hour. Slack, Twitter, email. A tap on the shoulder to come look at something. An impromptu meeting. All these distractions put your engineers in the mindset of being reactive and makes them feel incredibly unproductive.

Think about how you feel after days like that that are filled with random to-dos. You feel busy, tired, and like you did absolutely nothing — my least favorite feelings. If I keep days like this up for too long, I start to feel as if I'm not contributing to my team, not growing. I feel unfulfilled.

Busy work should be eliminated from your engineers' schedules as much as humanly possible. Engineers who are allowed to think in a free, fluid, and proactive manner build better software.

Important and urgent

Crises of any kind fall under the category of important and urgent. If an outage is impacting customers, the issue is both time sensitive and critical to your business. A crisis will always require immediate attention while having an impact on your long-term mission.

In addition to crises, certain deadlines fall into this category. I want to warn you away from manufactured crises. If you set a soft deadline or pick an arbitrary date by which you plan to complete something, the approach of that date is not a crisis. Occasionally, though, a deadline can be both urgent and important. If you have an annual conference and plan to release a set of new features, that deadline is extremely time sensitive and important.

Here are some additional examples of areas with deadlines that are both important and urgent:

- Potential partnerships
- Hiring
- Financial reporting
- Annual raises and rewards
- Personal emergencies

Usually you have an idea of when deadlines will become urgent, especially if they're an annual occurrence. Be sure to plan ahead so that you're not caught off guard or unnecessarily stressed. Emergencies happen, both to the business and its employees. Ideally, you'll move anything expected to the second quadrant.

Important, not urgent

Important but not urgent tasks are vital to the long-term health of your employees, products, and organization. These items are mission critical but lack specific deadlines.

In DevOps, these tasks include planning, continuing education, paying down technical debt, and strengthening team trust.

Additional examples of important but not urgent tasks are

- Building relationships
- Long-term product planning
- Practicing new skills
- Reducing bottlenecks
- Practicing failure
- Training
- Reading

The tasks that dwell in this quadrant are some of the most likely to be dropped. Because the urgency doesn't exist, people put them off indefinitely. Even when the tasks themselves could make you more effective for the things you do that are urgent, it's difficult to complete a task without having someone standing over your shoulder.

Even as I write this book, I depend on my editor to send me occasional emails as a reminder that I need to deliver chapters. This encourages me to write, even when I don't want to. The lack of this time-sensitive accountability is what leaves the items in this quadrant at risk of being forgotten.

Here are a few things to think about when considering important tasks without a specific deadline:

- **Clean code is easier to maintain.** It doesn't matter whether it's application or infrastructure code. Reducing technical debt by refactoring code or simplifying a process will quickly pay dividends to your team's overall velocity.

- **Engineers who trust each other are more effective.** Taking time to build rapport as a team and allowing relationships to form will create a more fluid process in the future. Trusting relationships remove friction, which in turn makes accomplishing difficult tasks easier.
- **Continual learning ignites neurons.** The worst thing you could do for your team is to make your engineers commodities — empty code monkeys who simply pump out work 40 hours a week. Instead, you want to create an environment in which engineers can continually push their skills, learn new technologies, and creatively solve problems.
- **Planning creates a road map.** Even if you end up having to adjust, pivot, or abandon a plan altogether, it will serve you to build one. Having a plan creates a vision for what's coming next, which allows people to ruminate and prepare. The discussions around planning are absolutely priceless in a DevOps organization because they can spur new ideas, important discussions, and creative problem-solving.

Perhaps the greatest challenge around this quadrant's items is that you don't know what's important. You can easily to fall into the habit of putting out fires. Checking email. Looking at Twitter. Responding to whatever is most pressing at the moment.

It's much harder — and takes a lot more discipline — to be aware of the things that need to get done that will improve your team's overall performance. Although delivering features is important and urgent, paying down technical debt is absolutely critical to building a healthy DevOps organization.

Leadership must have a clear vision of what's important for your organization. Remember, the *how* is not nearly as important as the *what*. If you clearly communicate what's important to everyone in the company, they can prioritize the work that best suits those goals.

One of the best strategies I have to combat the power of urgency over importance is scheduling time for email, Slack, and Twitter. This strategy applies to anything. Slack and Twitter are the big offenders for me; yours could be different. Recognize what continually prioritizes "urgent" tasks for you and create systems to prevent the reactive nature of that work.

Limit yourself to checking email two to three times a day at set times. Do the same with social media or chat applications. Let your colleagues, employees, and managers know that you do your best work when left uninterrupted and you therefore check these applications at specific times during the day. If they need you, they can call you. Also, your company should have a humane on-call rotation that allows for breaks and time away from being responsive to incidents. That rotation frees time for you to focus on long-term planning, continual education, and other team priorities.

Urgent, not important

The urgent but not important tasks are perhaps the most dangerous to your mission. They require immediate attention but don't help you achieve any of your team's long-term goals. In fact, spending time on these tasks may cause you to ignore tasks that are important to the overall health and velocity of your organization.

Interruptions that fall into this quadrant of the matrix include:

- Unscheduled interruptions
- Getting tapped on the shoulder
- Phone calls
- Some meetings
- Last-minute meetups

The hardest challenge around these tasks is they *feel* important. Knowing the distinction between what feels important and what is important is honed by experience and dedicated practice.

Ask yourself the following questions:

- Does this need to be done?
- Does this need to be done right now?
- Does this need to be done by me right now?

If you can defer a task or delegate it — without simply pushing the stone downhill — do it. Keep in mind the time frame when you do your best work. I try to leave the urgent but not important tasks for mid-to-late afternoon. I do my best work in the morning, especially if the work is creative in nature or particularly challenging. In the afternoon, I'm still around but I try to schedule rote tasks that need less brain to compute. Afternoon coffee meetings are a great way to handle the tasks that often find themselves in this quadrant of urgent but not important tasks.

TIP

Be wary of adding work to your engineers' plates that isn't important but seems urgent. Sometimes this decision has more to do with understanding and holding to the organization's priorities. I've seen many teams agree to seemingly small tasks to appease customers or high-paying clients. Be careful of making this type of work a habit. It does not serve the greater mission of the organization, plus it distracts your employees from other work and often does not appease the client to the extent you think it will.

One way of checking in on how much urgent but not important tasks are interfering with your goals is to regularly review your quarterly goals and key performance indicators (KPIs). Every Friday, for example, you might sit down and reflect on how this week contributed to the goals you have roughly 12 weeks to accomplish. What went well? What went poorly? What actions were neutral?

This regular self-reflection requires you to be diligent about tracking your work. If you don't have something like a ticket system for tracking work, consider using your calendar or — my favorite — a simple notebook to jot down what you worked on and when.

Neither important nor urgent

These tasks are the SQUIRREL! moments that distract people from their work. They are neither important nor urgent. This mindless activity is typically getting lost in the Internet. It could be scrolling through Instagram with no purpose. Or watching TV. Or getting lost on Reddit. Whatever the activity, it provides you with no personal or professional value.

My solution? Schedule time for these tasks. Seriously. I love watching lousy reality TV. *Real Housewives* is my (admittedly ridiculous) happy place. Now, I don't watch it during work hours, but because it serves as neither important nor urgent to even my personal goals, I schedule time to enjoy my mindless TV show. I take a bath once or twice a week and watch that week's episode.

Increasing Velocity

Velocity is one of those popular tech words that get thrown around a lot by "thought leaders." You may hear it frequently but not know exactly to what it refers. Its roots, like much of DevOps, are found in Agile software development.

Velocity is a measure used in sprint planning. Put simply, if you track your team's performance over a number of sprints, you can (within reason) predict the velocity of work in the coming sprints. Predicting velocity improves planning because you can roughly sketch out how much your team will be able to accomplish over *n* sprints.

In reality, I've seen few teams track performance in a way that allows for a predictable velocity. In fact, using it as a predictor is problematic for a number of reasons. For the reasons I explain in the following list, I encourage you to think of velocity as a single data point. Avoid using it as a single measure of your team's performance. If you put too much emphasis on velocity, you miss the other qualities and data that give you insight into the areas in which your team thrives and the areas in which it can improve.

- **It's impossible to "size" work.** Any sizing done in Agile — approximating how long completing a specific task will take — is an estimate. You should double or triple that estimate before even suggesting a deadline to an executive or stakeholder.

  As an engineer, I've sized stories thinking they would take days, only to have them turn out to be much easier than expected and only take a few hours. Likewise, I've estimated work to take only half a day and ended up in a rabbit hole of nested problems that took weeks to untangle. And it's not just me. Sizing is an industry-wide challenge. Because the sizing doesn't match up, measuring velocity based on stories completed gets tricky.

- **Team performance is more than speed.** A team of engineers can crank out dozens of features within a week or two. But the code will be a poorly tested mess of spaghetti code that is so impossibly complicated and sloppy that refactoring it would take longer than simply rewriting the original work. When you increase speed without automation, quality often suffers.

Increasing velocity requires optimizing your team performance while respecting the constraints and context your team experiences daily. Every engineering project must be completed in the constraints of your particular team and organization. I discuss the constraints of scope, deadline, quality, and budget in Chapter 7. A way to visualize these common constraints is with a triangle whose three boundaries represent speed, quality, and cost, as shown in Figure 12-2. Generally, you can choose two of the three. Quick, high-quality work will be expensive. Inexpensive, quick work will be poor quality. And inexpensive, high-quality work will likely be slow.

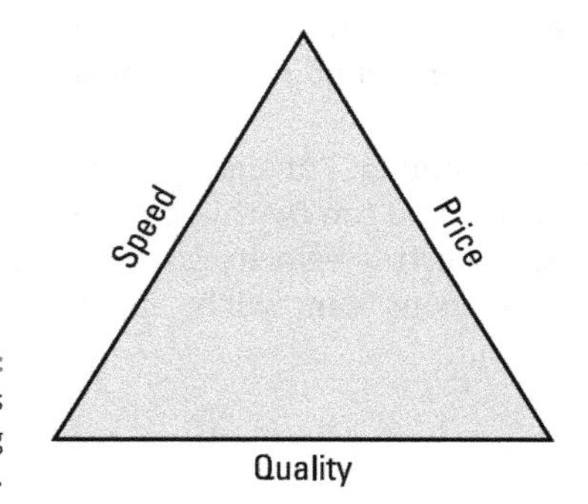

FIGURE 12-2: Three boundaries of engineering work.

You calculate velocity using two data points: unit of work and interval. The unit of work is simply what gets accomplished. You can use engineer hours (my preference) or something more abstract like Agile story points. Interval is the time duration.

Agile story points are arbitrarily assigned values that serve as a way of each team to create shared understanding. Teams typically use t-shirt sizing (extra small, small, medium, large, extra large) or the Fibonacci sequence (1, 2, 3, 5, 8, 13, 21). The sizes are relative to the others in the sequence. For example, a story sized as a 2 will take double the effort of a story sized as a 1. However, sizing never correlates cleanly to developer hours. You should never use story points as a way of comparing teams across the company because what constitutes a particular size will vary from team to team.

Sizing is beneficial because it gives engineers and product managers a way of talking about the developer resources required to accomplish a particular feature or bug fix. Engineers size work while keeping in mind the complexity of the work (or the area of the codebase that requires updating), the uncertainty around the work (engineers need time to figure out how to execute more verbose tasks), and the estimate on the time required to complete the work.

Take, for example, a week-long Agile sprint during which your engineering team plans to complete 32 story points. Now imagine that because of unexpected speed bumps, the team accomplished 27 story points' worth of work. For that week, the team velocity was 27 — the value of the story points associated with completed tickets.

You can begin to measure velocity over time by tracking the velocity week over week. Following is an example of how velocity can vary week to week, typically as a result of unexpected complexity in completing large tickets. Although Sprint 4 saw a dip in story points completed, the velocity over time stays roughly the same.

Sprint 1: 32 story points

Sprint 2: 28 story points

Sprint 3: 30 story points

Sprint 4: 14 story points

Calculation: (32 + 28 + 30 + 14) / 4 = 26

Velocity: 26

Although you should never use velocity as a way of comparing engineering teams across an organization (and thus never report it up the chain to executives who will do just that), it can serve as a baseline through which you can measure how DevOps practices improve your team productivity.

LOSING MONEY FAST

On August 1, 2012, an engineer forgot to replicate new code onto one of eight production servers at Knight Capital. Because of the speed of high-frequency trading, that one mistake caused the company to lose $440 million in less than an hour. That's not what I mean when I say fail fast. I love the Knight Capital fiasco as an example of why DevOps is vital to high-velocity organizations:

- **It involved human error.** You might say that one cause of the issue was that one "dumb" engineer who should have done their job better. The thing is, humans make mistakes. Humans are actually better at making mistakes than doing anything else. The systems you create in your DevOps organization must take that fact into account and work toward preventing human error — creating checks and redundancies to reduce the possibility.
- **The incident happened fast.** It happened so fast that by the time the team realized something was wrong, identified the issue, and fixed it, the damage was done. The software executed more than 4 million trades during the incident. The company lost nearly a third of its market value. As a result, the stock price tanked and the company had to raise $400 million a few days later to stay solvent.
- **Poor decisions led to cascading impacts.** The New York Stock Exchange (NYSE) received Securities and Exchange Commission (SEC) approval for a dark pool called the Retail Liquidity Program (RLP) in June 2012. The RLP would launch on August 1, 2012, which gave Knight Capital just over 30 days to prepare. The company developed the software in a scramble. Dead (unused) code — which was never intended for a production environment — was left in the system. They repurposed a flag to activate the RLP code rather than the dead code. The repurposed flag and unused code was the ultimate cause of the poor trade executions.
- **It highlights the need for automation.** A single engineer manually deployed the new code. No one conducted a review process. They had no automated verification in place to ensure that the correct builds were released to each server.
- **Initial alerting failed to notify engineers.** An hour and a half prior to initial trading, the system sent 97 emails with a vague error report to Knight Capital employees. But, as is apparent, email is a terrible vector for alerting. People don't prioritize email and generally don't open it in a timely manner. Despite the system's warnings, engineers did not take action.

This scenario is a nightmare. Seriously. Engineers wake up sweating just thinking about a technical Armageddon like the one Knight Capital endured. Preventing this doomsday scenario — along with hundreds of significantly smaller incidents — is one of the great benefits of DevOps.

High-frequency trading is relatively new and explosively fast, but the problems it presents aren't new, which is why DevOps has become such an important solution for many organizations. The industry as a whole has recognized the problems engineers face on a daily basis and, through DevOps, attempts to mitigate those challenges.

Improving Performance

Improving engineering performance can have sweeping impacts on the entire business. Streamlining the development life cycle and removing bottlenecks will serve to accelerate the overall performance of the business — ultimately increasing the bottom line. And if you think, as an engineer, that you shouldn't have to care about the business performance, you're wrong.

According to DevOps Research and Assessment (DORA), high-performing teams consistently outpace their competitors in four key areas:

- **Deployment frequency:** This term refers to how often your engineers can deploy code. Improving performance aligns with deploying multiple times per day as desired.
- **Lead time:** Lead time is how long you take to go from committing new code to running that code in a production environment. The highest performers, according to DORA, have a lead time of under an hour, whereas average performers need up to a month.
- **MTTR (Mean Time to Recover):** MTTR refers to how long you take to restore a service after an incident or outage occurs. Ideally, you want to aim for under an hour. An outage costs serious money, especially when it impacts profit centers of the application. Long outages destroy trust, decrease morale, and imply additional organizational challenges.
- **Change failure:** This term refers to the rate at which changes to your system negatively impact the performance. Although you will never reach a change failure rate of zero percent, you can absolutely approach zero by increasing your automated tests and relying on a deployment pipeline with continuous integration checks and gates — all of which ensure quality.

Eliminating perfection

I believe strongly in the mantra "Done is better than perfect." It seems to be one of these impossible-to-attribute quotations, but the words nonetheless speak truth. Attempting to attain perfection is an enemy of effectiveness and productivity. I think most engineers suffer from some version of analysis-paralysis — a mental affliction that limits your productivity in an attempt to overanalyze your work and sidestep any potential mishap.

Training imperfection into your work requires you to embrace the possibility of failure and the inevitability of refactoring. In Chapter 13, I talk about creating feedback loops around the customer and looping back to various stages of the pipeline. In Part 2 of this book, I dedicate a chapter to each phase of the software development pipeline in a linear flow. Here, you're connecting the ends to bend the line into a circle.

When you think iteratively and circularly, pushing out code that's not perfect seems a lot less scary because the code isn't carved into stone. Instead, it's in a temporary state that you improve frequently as you gather more data and feedback.

Designing small teams

You've likely heard of Amazon's "two-pizza" teams. The concept broadly speaks to the importance of small-sized teams. Now, the exact number of people that comprise a two-pizza team varies according to your appetites.

I grew up Methodist, and one of the things the Methodist church emphasizes is small groups. All small groups are kept under 12 people — the number of original disciples. I tend to keep that principle in mind even now. When a group approaches 9, 10, or 11 people, I split it into two. I find that the sweet spot for group size is around 4–6 people. Your exact number may vary depending on the people involved, but the point is this: When groups get too large, communication becomes challenging, cliques form, and the teamwork suffers.

I've added one other bonus goal when forming teams: even numbers. I believe strongly that people need a "buddy" at work — someone they can trust above all others. In even-numbered groups, everyone has a buddy and no one is left out. You can pair off evenly and it tends to work well. Forming even-numbered groups isn't always achievable because of personnel numbers, but it's something to keep in mind.

A formula for measuring communication channels is $n (n - 1) / 2$, where n represents the number of people. You can estimate how complex your team's communication will be by doing a simple calculation. For example, the formula for a two-pizza team of 10 would be $10 (10 - 1) / 2 = 45$ communication channels. You can imagine how complex larger teams can become.

Tracking your work

If you can get over the small overhead of jotting down what you do every day, the outcomes will provide you with exceptional value. Having real data on how you use your time assists you in tracking you and your team's efficacy. As Peter Drucker famously said, "If you can't measure it, you can't improve it."

How many days do you leave work feeling like you did nothing? You just had meeting after meeting or random interruptions all day. I have the same problem. I'm fairly terrible at tracking my time, and when I'm not disciplined about writing down what I do each day, I can quickly feel much less effective than I actually am. The divergence between our feelings of efficacy and the reality of our efficacy is dangerous territory for any team.

I encourage you to use pen and paper rather than some automated tool for this. Yes, you can use software to track how you use your time on your computer. It can tell you when you're reading email, when you're slacking, and when you're coding, but it lacks nuance and often misses or incorrectly categorizes large chunks of time.

After you have an idea of what you're doing and when, you can start to identify which activities fall into which quadrants of the Eisenhower Decision Matrix. What busy work are you doing routinely that provides no value to you or the organization?

Reducing friction

One of the best things a manager can do for an engineering team is to leave them alone. Hire curious engineers who are capable of solving problems independently and then let them do their job. The more you can reduce the friction that slows their engineering work, the more effective your team will be. Reducing friction includes the friction that exists between teams — especially operations and development. Don't forget specialists like security, either.

Aligning goals and incentives increases velocity. If everyone is focused on achieving the same things, they can join together as a team and move methodically toward those goals. Flip to Chapter 15 to read more about aligning incentives.

Humanizing alerting

Every engineering team has alerts on actions or events that don't matter. Having all those alerts desensitizes engineers to the truly important alerts. Earlier in this chapter, I describe how Knight Capital got into trouble by ignoring 97 emails from the system. I'd venture a guess that they had become conditioned to ignore email alerts because of an overabundance of messages. Alert fatigue ails many engineering organizations and comes at a high cost. If you're inundated daily, picking out the important from a sea of the unimportant is impossible. You could even say that these messages are urgent but not important . . . .

WARNING

Email is not an ideal vehicle for alerting because it's not time sensitive (many people check email only a few times a day) and it's easily buried in other minutiae.

Applying what you've learned about rapid iteration, reevaluate your alerting thresholds regularly to ensure an appropriate amount of coverage without too many false positives. Identifying which alerts aren't necessary takes time and work. And it'll probably be a little scary, right? Deleting an alert or increasing a threshold always comes with a bit of risk. What if the alert is actually important? If it is, you'll figure it out, I promise. Remember, you can't fear failure in a DevOps organization. You must embrace it so that you can push forward and continuously improve. If you let fear guide your decisions, you stagnate — as an engineer and as an organization.

» Collecting and communicating feedback

» Accelerating your iteration

Chapter 13
Creating Feedback Loops around the Customer

I believe that the age of the MBA CEO is over and that engineers represent the next generation of company leadership.

These claims may seem bold, but hear me out. Every company is adopting technology to remain relevant, improve services, and compete for customer attention.

Who understands tech more than an engineer? Engineers understand the tiny details that, together, make up the larger picture. But this engineers-as-leaders future requires engineers to appreciate two aspects that they used to be able to ignore: the mission and business of the greater organization; and the importance of customer experience and feedback.

In Chapter 6, I talk about why the mission of your business is important for DevOps organizations and how inviting other areas of the business to participate in collaborative planning benefits your engineering team. In this chapter, I talk about the importance of customer experience and feedback, including how to create a customer feedback process so that you can you begin to integrate that feedback into your software development process. In this chapter, you find out how to create a feedback loop, collect customer feedback, and accelerate your iteration through continual feedback.

Creating a Customer Feedback Process

Build-Measure-Learn is a concept introduced by Eric Ries in his book *The Lean Startup*. Customer feedback falls into the loop depicted in Figure 13-1. You want to build a prototype, collect data to measure success, and learn from the failures. This never-ending learning process makes way for numerous iterations of your product. And I mean what I say. The process never ends. You don't reach a destination where you may hang up your hat and announce you're finished. You must constantly learn, adapt, and refine.

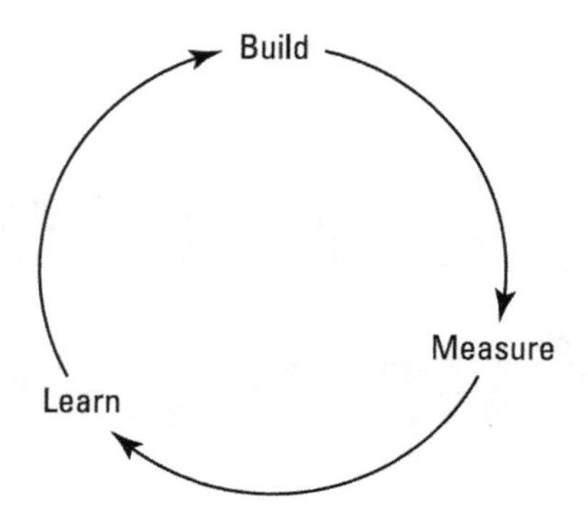

FIGURE 13-1: Build-Measure-Learn.

This idea of continual learning and improvement is one of the most important concepts for a DevOps organization and is often overlooked. Continuous integration, continuous delivery, and continuous deployment means nothing if those decisions aren't informed by actual customer feedback. Thus, continuous feedback is central to your DevOps practice.

Customer feedback has three main purposes:

- **Retaining customers:** Keeping a customer is significantly cheaper than acquiring new users.
- **Empowering employees:** In DevOps, people talk a lot about ownership and accountability, although not in a punitive sense. Instead, these concepts refer to giving your engineers a sense of pride in their work. You give them the power to gather and act on customer feedback.
- **Improving products:** No better way exists to iterate on your assumptions than to ask the people who use your products (and, you hope, pay you). Listening to honest feedback is a characteristic of any high-performing team.

I work in developer relations, and feedback is the most important aspect of my job as an advocate at Microsoft. I talk to engineers and listen to community feedback. Then I deliver that same feedback to the appropriate Microsoft product teams.

I believe strongly in listening to customers. No, the customer is not always right. But they will always give you useful information. Whether you act on it is a separate consideration, but collecting it is one of the most useful actions you can take — as an engineer and especially as a manager.

Creating a Feedback Loop

In this section, I expand on Reis's initial concept as well as shift the startup philosophy of Build-Measure-Learn to the feedback philosophy of Receive-Analyze-Communicate-Change, as shown in Figure 13-2.

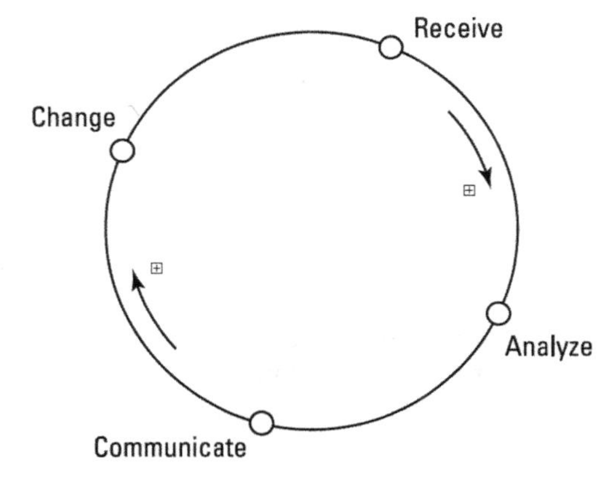

FIGURE 13-2: Receive-Analyze-Communicate-Change.

Receive

I touch on a few ways you can receive feedback in the "Collecting Feedback" section, later in this chapter, but here I want to emphasize one thing: There is no one way to collect feedback. In fact, opening your organization to multiple ways of receiving feedback is the most helpful and thorough process.

You have many ways to receive feedback via unofficial channels. A friend might mention that a feature seems slow or doesn't quite work as expected. A beta user might send a random email. Someone might stop one of your engineers at a conference to let them know they love (or hate) a particular product.

However you receive feedback, you need to create a process through which anyone — truly anyone in your organization, from the engineers to the executives — can receive and pass along feedback. If you don't have a particular process in mind, start with a simple feedback form. It should collect in one place, where one person can own it. The shorter, the better. It doesn't need to be exhaustive. Shorter forms have a higher likelihood of actually being filled out.

Potential questions could include:

- **Name (and contact information):** Who is submitting this feedback? Ensure that you have a way of following up.
- **Method:** How was this feedback conveyed? Tracking the method gives you information about how customers usually reach out to you and your employees.
- **Feedback:** Leave this part as open-ended as possible. You can categorize and analyze the feedback in the next phase.

Analyze

This phase of your feedback loop is critical and should be left to someone who feels passionate about the topic, is close to the tech, and feels a sense of ownership over the product. Because feedback in its most raw state is often given relatively free-form, you categorize and analyze the data in the analyze phase.

Start to create categories in which you can group pieces of feedback. You can group items by feature or quality (for example, slow). Set up the categories so that you can quickly discern how many pieces of feedback you receive per category. Also a good idea is to identify which features on your website are most critical to your business and which are most often used by customers. All these efforts help you to prioritize work as you proceed.

Finally, make sure to keep this process relatively open. Although one person should own the analyze phase, provide as much sunlight to the process as possible. Ensure that everyone can access documents and that anyone can join the feedback process if they're interested. This open access is especially useful when onboarding new employees. There are few better ways of introducing your product to new colleagues than by having them read customer feedback.

Communicate

The third phase of your customer feedback process must include communicating the feedback to your engineering and product teams. This phase can be even more tricky than collecting the feedback in the first place. This communication phase is where most organizations stumble.

How you communicate the feedback determines its impact. You should never approach an engineering group or product team with a sense of condescension. You are not smarter than they are. You are not more important than they are — even if you're the CEO. Engineers and product managers make hard decisions within (sometimes extremely tight) constraints with the information they have at the time. Have some empathy and respect for their work.

As hostile as some customers may be, you can feel assured that you will likely never have to face their wrath again. Colleagues, on the other hand, are a bit more tricky, and effectively communicating negative feedback requires extreme trust.

Always communicate feedback gently. Although you don't need to sugarcoat negative feedback, be careful of your attitude — even your body language — when conveying it. Also be sure to pass along positive feedback as quickly as you do negative. Not only will it make everyone feel good about their work, it will also help them determine the correct direction for their team and their product.

Engineers, especially developers, tend to view their code as their babies — even if those babies are particularly ugly. Telling an engineer that their baby is ugly isn't an easy or fun process. It's really hard.

Communicating feedback to engineers requires striking a balance between positive and negative reviews of a product. It also requires an acknowledgement that creating any product is really, really, really hard. The humans behind mediocre to terrible products are still humans who are doing their best. That work and effort deserves respect.

Change

Implementing change in a technical product comes down to having clear priorities. What is the vision for your product? Which services are most critical to maintaining your current customers? What about reaching new users? What is the core problem your product solves?

Just because one user complains that a service is slow doesn't mean that it is. Nor does it mean that you should change anything — even if it is slow. What are your performance standards? Is this a feature many people use?

This sense of priority is why I'm a big fan of Kanban boards. (A Kanban board is a visual depiction of work in progress at various stages and the work coming up next.) Continually ranking and tracking work helps engineers do their jobs more easily. Ranking and tracking remove some of the pressure of figuring out what work is most important to do next or will give them the best shot at a promotion.

Finally, be cautious of how quickly you change. Yes, you have moments in your business when you need to react, and fast. But those moments are few and far between. Change should be a measured decision, with customer feedback acting as one piece of data.

Collecting Feedback

You have many options for collecting feedback from customers, and I discuss the pros and cons of each option in the following sections. Consider each option thoughtfully and choose the best plan for your organization. Often, casting a wide net and using a combination of methods provide the most diverse and holistic viewpoints.

Satisfaction surveys

The first option that likely comes to mind is a customer survey. Send your customers a series of prepopulated questions via email and collect the responses.

Surveys are perhaps the lowest-hanging fruit when it comes to collecting feedback. They require relatively little work, and you can send them to nearly every customer and package the results in neat little bundles for executives and other stakeholders.

You can measure customer satisfaction via a survey in a relatively empirical manner. Though the opinions are just that, the scores are more quantitative in nature, allowing for nearly limitless ways to analyze the data.

However, can you think of a worse way to spend your time than to fill out a detailed and yet oh-so-sterilized 30-question survey? I loathe surveys, and I doubt I'm alone.

Surveys are time intensive to answer. Sure, some of your customers may take their time and answer thoughtfully. The majority, though, will Christmas-tree the answers, popping ratings randomly, or mark the same answer for each question.

Although the price of online surveys ranges from free to relatively cheap, the vendors that offer to conduct more formal surveys and analyze the results are budget-breakingly expensive.

Finally, determining whether you're asking the right questions is extremely difficult. Because surveys produce a single response with ranked opinions, digging into the actual problems is a challenge. Do you know how your customers use your product? Perhaps they're satisfied with the product because it's the best option on the market, but far from ideal. What about your service can customers not live without? What features are just so-so? These questions are hard to answer via a survey. Here are the pros and cons of a survey:

Pros	Cons
Easy to produce	Takes a long time to answer
Quantitative data is simple to package and analyze	Answers may contain misleading data

Ultimately, I think surveys are useful to measure a customer's satisfaction with small interactions. Did your user feel that a customer support representative helped them? Did your user feel that documentation fully explained how to use an API?

Most products require more complex and multifaceted feedback vehicles to gather detailed quantitative and qualitative data around your customer's satisfaction with your product.

Case studies

Another option for collecting feedback is to bring in big-spending customers and make a day of it. You and your sales team can wine and dine them a bit, thank them for being customers, and then ask questions that instigate healthy conversation around the product. An event like this provides ample opportunities to allow the customer to upgrade their service or hear about other offerings that could benefit them (and your bottom line).

In addition to offering potentially fruitful conversations, case study meetings create — with client permission — great marketing stories. You get a better idea of how a high-paying customer uses your product, thus allowing you to potentially replicate that type of client.

One of the major flaws with case study meetings relates to the people involved. You're unlikely to invite your $5-per-month customers into the office to talk in depth on how they use the product. But those customers — in aggregate — make up a sizable chunk of your business. Don't underestimate the importance of market share and the power of a large number of people using your product, especially if they speak about it favorably.

In case study meetings, you speak almost exclusively to enterprise clients, and the people outside engineering, specifically executives and salespeople, dominate the conversations. This situation reduces or eliminates the opportunities for engineers to dive deep with a customer's engineers and talk tech. Here are the pros and cons of case studies meetings:

Pros	Cons
Great for customer face time	Likely limited to enterprise customers
Provide more in-depth conversation	Typically involve executives over engineers
Can potentially sell additional services	Travel costs add up
	Difficult to package, analyze, and communicate

Dogfooding

Perhaps the easiest way to receive feedback is to use the product yourself! That's what dogfooding means. Many companies utilize their own (often developer-centric) products internally. Dogfooding is simply an industry term to describe the practice of doing just that. For example, at Microsoft, all of Azure is hosted on Azure, which means that Microsoft is invested in the future of its product. Employees are the first to get access to new features and services, which gives Microsoft employees key insights into which areas need improvement.

Dogfooding has two qualities that you'll struggle to find in other methods of feedback:

- **It's close to home.** If you have a healthy culture, your employees are more likely to feel free to provide harsh feedback and initiate product changes internally — often long before customers complain.
- **It's faster than other methods.** Dogfooding acts as a mixture of QA testing and beta testing. Typically, when you dogfood a product, you use a preview mode of your own product.

You can accomplish a preview of your product through certain deployment processes (see Chapter 11 for details on deployment) or by utilizing feature flags to determine which customers see which features. For example, if you put all your employees under a particular set of permissions — likely grouped into the role "employee" — you can open the preview features to employees without revealing them to your average user.

In addition to allowing your employees to dogfood your product, I highly recommend opening this preview to customers who select to preview or beta test advance features. These users are typically extreme early adopters who like to play at the absolute bleeding edge of technology. Not only will they provide you with extremely valuable feedback earlier in your development cycle, they may also become evangelists for your product — turning more users into customers.

Here are the pros and cons of the dogfooding method of collecting feedback:

Pros	Cons
Faster and cheaper than customer feedback	Can lack diversity of thought
More likely to provide honest feedback	Requires some redundancy to reduce the risk to your business
Contained to a small set of users	

The speed at which you can iterate on dogfood feedback will vary based on your internal processes. But I've highlighted the extreme examples in Figure 13-3, which shows a customer feedback pipeline, and Figure 13-4, which shows a dogfooding feedback pipeline.

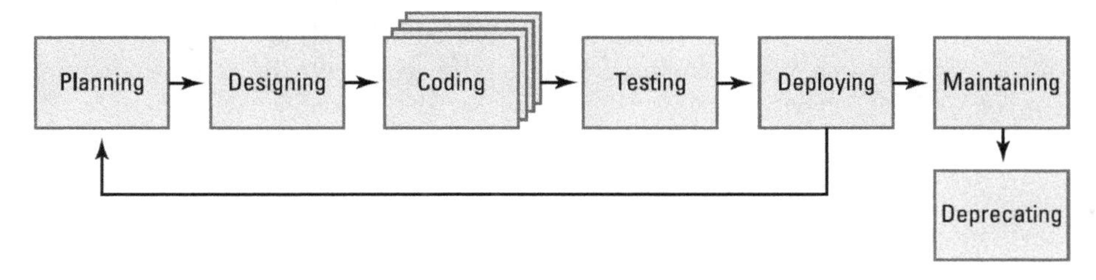

FIGURE 13-3: Customer feedback pipeline.

With customer feedback, you have to wait until your product is fully deployed to production to hear from customers. You might wait for days or weeks before you get the first piece of feedback. Then you must take that feedback, put it through your analysis process, and start planning to integrate any necessary changes into the next appropriate sprint or release.

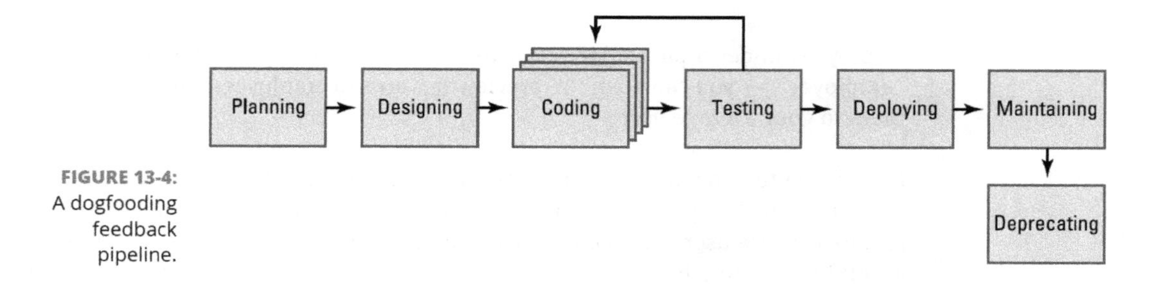

FIGURE 13-4: A dogfooding feedback pipeline.

If you dogfood your own product, the feedback cycle is condensed and you can potentially iterate much more quickly. The example shown in Figure 13-4, in which you immediately return to coding after receiving feedback, is extreme — you may instead choose to put the feedback into the planning process and proceed accordingly. But if just a small change is needed — perhaps a button is displaying oddly, or an engineer finds a typo in the text — engineers can feel empowered to make small, inconsequential decisions freely and immediately.

Small changes are extremely quick to implement, can make engineers feel more involved and accountable to the product, and can avoid potentially embarrassing mishaps with customers — especially if the bug isn't deployed to users yet.

Asking for Continual Feedback

From a DevOps perspective, continual improvement requires continual feedback. Your organization would be wise to build feedback loops around your customers as you transform your engineering culture and development life cycle with DevOps.

One of the goals of continual feedback is to make receiving feedback a daily process. Achieving this goal might not happen overnight. If you're currently conducting an annual survey, aim for monthly contact with customers. Then make it weekly and keep increasing the frequency until you reach the point where you talk to customers so frequently that it's second nature.

In addition to collecting continual feedback, letting your customer know that their feedback had an impact on the product is an incredibly rewarding experience — for everyone.

Net promoter score (NPS)

Created by Fred Reichheld, the concept of the net promoter score puts customers into three groups:

- » Promoters
- » Passives
- » Detractors

The major benefit of NPS is its simplicity. You ask your customers a single question: "How likely would you be to recommend this product to a friend?" The scores categorize your customers as follows:

- » **Promoters:** Scores of 9 or 10
- » **Passives:** Scores of 7 or 8
- » **Detractors:** Scores of 6 and under

The follow-up is perhaps the most important aspect of NPS. You must ask your customers why they gave the score that they did. That qualitative information is absolute gold and should inform your decisions moving forward on product improvements, service deprecations, and new-feature planning.

Consider how many interactions your users have with your product. Those interactions may be monthly, weekly, or daily. Each time a customer logs in, you have another opportunity to turn that customer into a promoter. A customer who has become an evangelist is, just like the internal DevOps evangelists discussed earlier in this book, incredibly valuable in persuading others.

Finding a rhythm

In collecting customer feedback, the following steps are the ones I've found to be the most effective:

1. **Ask for feedback.**

   You can ask directly or indicate a general openness to feedback. The former strategy will collect more feedback from a wider audience. Either way, ensure that you make it easy for customers to get in touch with you. Add a chat dialog box to your website or list an email address that gets answered in a reasonable time frame.

2. **Listen and take notes.**

   One of the worst things you can do in a conversation with a customer is to shut them down, argue, or assume that they're wrong — even if you genuinely think they are. Saying "no" is the fastest way to end a productive conversation. Instead, embrace the fundamental phrase of improvement: "Yes, and. . ." Take notes so that can communicate the feedback faithfully to others later.

3. **Analyze the data.**

   It's great if you can collect quantitative data from the customer, but don't press them. Instead, learn to adapt the qualitative complaints into quantitative data points. For example, if someone says they hit API limits within minutes, you can deduce how many times they're likely making API requests based on your rate limiting for the specific API.

4. **Follow up.**

   If something isn't clear after speaking to the customer, or you can't replicate the problem, call them back. You're not bothering them; you're engaging with them. In these initial conversations, be sure not to give assurances of a fix — unless you're absolutely sure that you can provide that fix.

5. **Categorize and track.**

   Come up with specific categories into which you can group pieces of feedback. Sometimes a customer complaint is a one-off and isn't representative of your greater customer base. Here's the rule I use: "If one person says you have a tail, ignore them. If a hundred people say you have a tail, look behind you." Keeping track of how many people have complained or praised a particular feature gives you a better picture of not only which features are most used but also which are most problematic (or wonderful).

6. **Communicate with product teams.**

   If you don't have your actual engineers owning this feedback loop, you need to bring this information to them in a palatable way. Even if one of the engineers is the messenger, that individual still needs to convey the feedback and help the team decide what action, if any, is needed. It's vitally important to form trusting relationships before giving and receiving hard feedback. If you fail to invest in establishing rapport, it doesn't matter how much feedback you bring back to the team. It will fall on deaf ears.

7. **Rinse and repeat.**

   Meet frequently, and even if you have no tangible feedback to give, you should meet with the product team on a regular basis. The cadence will help you maintain collaboration as a regular and "normal" process. Meeting frequently prevents teams from feeling like they're in trouble if an impromptu meeting is called and allows you to learn from your teams. You can figure out what they're working on, see what they're passionate about, and learn what they want to see in the road map. Listen to them. This effort is a collaborative one and it's important to balance listening with talking.

These are the basic steps for collecting feedback, but I encourage you to experiment. Think through what you believe will work for your team, your organization, your product, and your customers.

In DevOps, everyone's accountable but not everyone is an owner. Be sure that one person "owns" customer feedback for a specific product or service. Ideally, if you've designed your engineering organization to have product teams, each team can collect feedback for the services they build. Even so, encourage one person to own the process. You don't give someone ownership so that you have a neck to wring. (Remember, you're creating a learning organization.) Instead, the idea is to empower the team to become fully autonomous in collecting feedback and implementing changes. When you release control and trust your engineers to be excellent, amazing things can happen.

» **Deciding on job titles**

» **Hiring and interviewing for DevOps roles**

Chapter 14

DevOps Isn't a Team (Except When It Is)

Forming teams that support your DevOps culture can be one of the trickier parts of your DevOps transformation. If your broader organization continues to encourage silos via misaligned goals and incentives, your team structure won't matter. You will struggle to get your DevOps approach off the ground.

In this chapter, I focus on how you set your team up for success. There are generally three ways of approaching team structure in a DevOps culture: aligning teams, dedicating teams, and creating cross-functional product teams. Each approach has advantages and disadvantages. This chapter delves into forming teams with DevOps in mind, along with recruiting, interviewing, deciding on job titles, and dealing with problematic employees.

Forming DevOps Teams

DevOps has no ideal organizational structure. Like everything in tech, the "right" answer concerning your company's structure depends on your unique situation: your current team, your plans for growth, your team's size, your team's available skill sets, your product, and on and on.

Aligning your team's vision should be your first mission. Only after you've removed the low-hanging fruit of obvious friction between people should you begin rearranging teams. Even then, allow some flexibility.

If you approach a reorganization with openness and flexibility, you send the message that you're willing to listen and give your team autonomy — a basic tenet of DevOps. You may already have a Python or Go developer who's passionate and curious about infrastructure and configuration management. Maybe that person can switch into a more ops-focused role in your new organization. Put yourself in that person's shoes. Wouldn't you be loyal to an organization that took a risk on you? Wouldn't you be excited to work hard? I certainly would. And that excitement is contagious. In the next few sections, I describe how to align the teams you already have in place, dedicate a team to DevOps practices, and create cross-functional teams — all approaches from which you can choose to orient your teams toward DevOps.

REMEMBER

You can choose one approach and allow it to evolve from there. Don't feel that this decision is permanent and unmovable. DevOps focuses on rapid iteration and continual improvement. That philosophy applies to teams as well.

Aligning functional teams

In this approach, you create strong collaboration between your traditional development and operations teams. The teams remain functional in nature — one focused on ops, one focused on code. But their incentives are aligned. They will grow to trust each other and work as two teams yoked together.

For smaller engineering organizations, aligning functional teams is a solid choice. Even as a first step, this alignment can reinforce the positive changes you've made so far. You typically start the alignment by taking the time to build rapport. Ensure that each person on both teams not only intellectually understands the other team's role and constraints but also empathizes with the pain points.

I recommend enforcing a policy of "You build it, you support it." This policy means that everyone — developer and operations person alike — participates in your on-call rotation. This participation allows developers to start understanding the frustrations of being called in the middle of the night and struggling while foggy-eyed and caffeine-deprived to fix a bug that's impacting customers. Operations folks also begin to trust your developers' commitment to their work. Even this small change builds an extraordinary amount of trust.

A word of caution: If developers fight hard against being on call, a larger problem is at play in your organization. The pushback is not uncommon because being on call is wildly different from their normal day-to-day responsibilities. The pushback often comes from a place of discomfort and fear. You can help mitigate this reaction by addressing the fact that your developers may not know what to do the first few times they're on call. They may not be familiar with the infrastructure, and that's okay. Encourage them to escalate the incident and page someone with more experience. Finally, create a runbook with common alerts and what actions to take. Providing this resource will help to assuage some fear until they begin to get the hang of things.

Another tactic to help spur collaboration is to introduce a day of shadowing, with each team "trading" a colleague. The traded person simply shadows someone else on the team, sits at their desk (or in their area), and assists in their day-to-day responsibilities. They may help with work, discuss problems as a team (pair programming), and learn more about the system from a different point of view. This style of teaching isn't prescriptive. Instead, it lends itself to curiosity and building trust. Colleagues should feel free to ask questions — even the "stupid" variety — and learn freely. No performance expectations exist. The time should be spent simply getting to know each other and appreciating each other's work. Any productive output is a bonus!

In this alignment approach, both teams absolutely must be involved in the planning, architecture, and development processes. They must share responsibilities and accountability throughout the entire development life cycle.

Dedicating a DevOps team

A dedicated DevOps team is more an evolution of the Sys Admin than a true DevOps team. It is an operations team with a mix of skill sets. Perhaps some engineers are familiar with configuration management, others IaC (infrastructure as code) and perhaps others are experts in containers or cloud native infrastructure or CI/CD (continuous integration and continuous delivery/development).

If you think that putting a group of humans into an official team is enough to break down silos, you're mistaken. Humans are more complex than spreadsheets. Hierarchy doesn't mean anything if your silos have entered a phase in which they are unhealthy and tribal. In toxic cultures, a strongman style of leadership can emerge that is almost always followed by people taking sides. If you see this on your own team, you have work to do.

Although any approach may work for your team, this dedicated team approach is the one I suggest you think through the most. The greatest disadvantage of a dedicated DevOps team is that it easily becomes a continuation of traditional engineering teams without acknowledging the need to align teams, reduce silos, and remove friction. The risks of continuing friction (or creating more) are high in this approach. Tread carefully to ensure you're choosing this team organization for a specific reason.

The benefits of this approach is having a dedicated team to address major infrastructure changes or adjustments. If you're struggling with operations-centered issues that are slowing down your deployments or causing site reliability concerns, this might be a good approach — even temporarily.

I also like a dedicated team if you're planning on moving a legacy application to the cloud. But rather than calling this team a DevOps team, I'd label it an automation team. This dedicated group of engineers can focus completely on ensuring that you've set up the correct infrastructure and automation tools. You can then proceed with confidence that your application will land in the cloud without major disruption. Still, this approach is temporary. If you keep the team isolated for too long, you risk going down a slippery slope from rapid growth to embedded silo.

Creating cross-functional product teams

A *cross-functional team* is a team formed around a single product focus. Rather than have separate teams for development, user interface and user experience (UI/UX), quality assurance (QA), and operations, you combine people from each of these teams.

A cross-functional team works best in medium to large organizations. You need enough developers and operations folks to fill in the positions of each product team. Each cross-functional team looks a bit different. I recommend having at a minimum one operations person per team. Do not ask an operations person to split their responsibilities between two teams. This scenario is unfair to them and will quickly create friction between the two product teams. Give your engineers the privilege of being able to focus and dig deep into their work.

REMEMBER

If you're organization is still small or in the startup phase, you can think of your entire engineering organization as a cross-functional team. Keep it small and focused. When you begin to approach having 10–12 people, start thinking about how you can reorganize engineers.

Figure 14-1 shows what your cross-functional teams could look like. But keep in mind that their composition varies from team to team and from organization to organization. Some products have a strong design focus, which means that you may have multiple designers in each team. Other products are technical ones designed for engineers who don't care much for aesthetics. Teams for that kind of product may have one designer — or none at all.

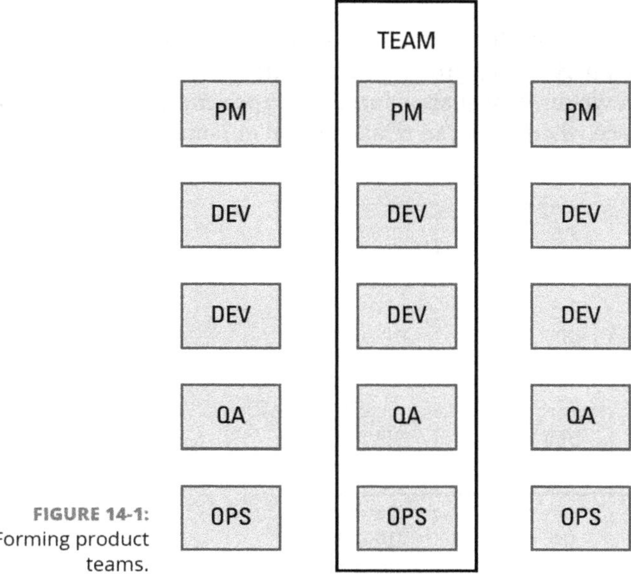

FIGURE 14-1: Forming product teams.

If your organization is large enough, you can certainly create multiple teams using the ideas and approaches described in this section of the chapter. Remember that your organization is unique. Feel empowered to make decisions based on your current circumstances and adjust from there. Here are some possible combinations of various types of product teams.

- **Legacy Product Team:** Project Manager (PM), Front-end Developer, Back-end Developer, Back-end Developer, Site Reliability Engineer (SRE), Automation Engineer, QA Tester
- **Cloud Transformation Team:** SRE, SRE, Operations Engineer, Automation Engineer, Back-end Developer
- **MVP Team:** PM, Designer, UX Engineer, Front-end Developer, Back-end Developer, Operations Engineer

The downside of a cross-functional product team is that engineers lose the camaraderie of engineers with their same skill sets and passions. Having a group of like-minded individuals with whom you can socialize and from whom you can learn is an important aspect of job satisfaction. I offer a solution to this issue in Figure 14-2.

As shown in the figure, you can give your engineers dedicated work time to spend with their tribes. You can do something as generous as paying for lunch once every week so that they can get together and talk. Or you might provide 10–20 percent of work time for them to work on projects as a tribe. Either way, you need your engineers to stay sharp. Tribes share industry knowledge, provide sound feedback, and support career growth. Provide time for your engineers to learn from people with whom they share education, experience, and goals. This time provides a safe place where they can relax and feel at home.

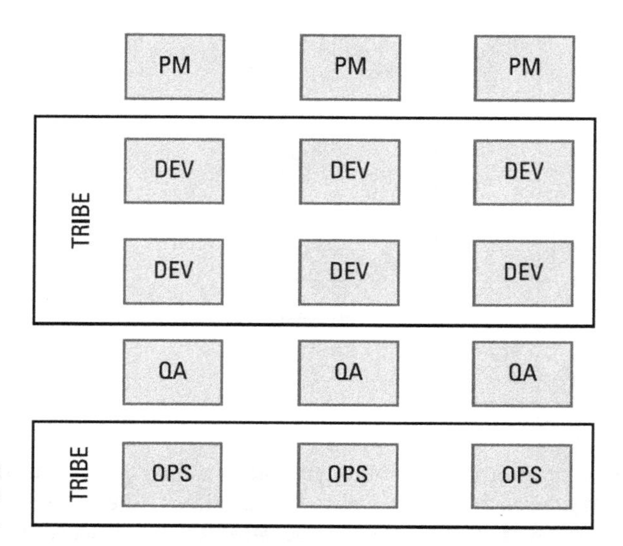

FIGURE 14-2: Making space for tribes.

No amount of perfect finagling will overcome the shortfalls of a bad organizational culture. But if you've paid attention so far and made the appropriate strides, the next step is to form teams that reinforce the cultural ideals you've already put in place.

Interviewing Quickly (But Not Too Quickly)

No matter how you organize your teams internally, you still need to hire people. Whether the reason for hiring is to expand your team or replace an engineer who moved on, hiring is always time intensive, expensive, and —face it — exhausting.

One of the challenges of this booming tech economy is that finding and hiring the best has become increasingly difficult. It's an engineer's market. Engineers of all kinds are in demand and every company faces a shortage of quality employees. The demand is even higher for the few magical engineers who are generalists, those who have wide-ranging experience and interests. These engineers are the ones you need most in a DevOps organization.

When you've found someone you want to hire, you need to move fast. Otherwise, you risk losing your newly found talent to competitors. But moving fast has its risks, including hiring someone who appeared wonderful and turned out to be disappointing. I cover how to handle hiring mistakes in the "Firing Fast" section, later in this chapter.

Deciding on a Job Title

DevOps is not a job title. It's a philosophy, a methodology, and an approach to removing friction in the software delivery life cycle. Yet, "DevOps Engineer" is an in-demand role at hundreds, if not thousands, of companies.

The war against DevOps as a job title has been lost, and the time has come to accept that fact. Adding DevOps to a title or role allows engineers to ask for $10,000–$15,000 more in annual salary (though I've seen as high as a $35,000 pay bump) as well as have a stronger negotiating position when interviewing for a new role. I would never cheat an engineer from utilizing every possible angle to progress toward their career goals.

When I set out to research common job titles for a DevOps-related engineering role, some of the results surprised me. I started at Google Trends and compared three roles:

- DevOps Engineer
- Release Engineer
- Site Reliability Engineer

You can see more data around this research at `https://g.co/trends/gZACs` as well as view the summary in Figure 14-3, which shows job titles that are associated with DevOps. DevOps Engineer is the clear winner, beating the others by a significant margin.

TIP

The role of Site Reliability Engineer is increasing in popularity. In many ways, SRE represents the evolution of DevOps and will continue to grow. I'll be shocked if SRE doesn't approach the popularity of DevOps Engineer as a job title over the next few years, if not surpass it. I suggest tracking this evolution closely.

The real surprise came when I added Automation Engineer to the list of job titles. In my anecdotal experience, Automation Engineer isn't a particularly popular job title. Yet, the results of my comparison via Google Trends contradict my initial belief. You can dig more into the data in Figure 14-4 and directly at `https://g.co/trends/5x6wY`.

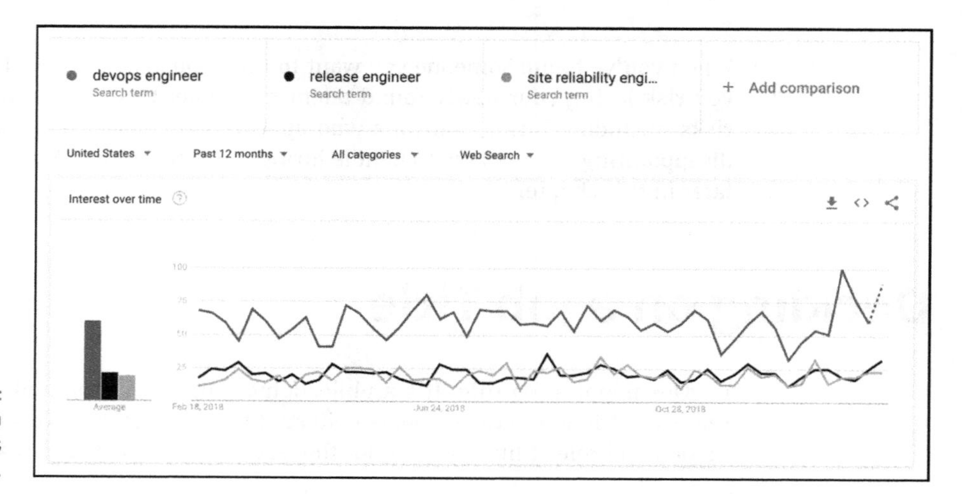

FIGURE 14-3: Common U.S. DevOps job titles.

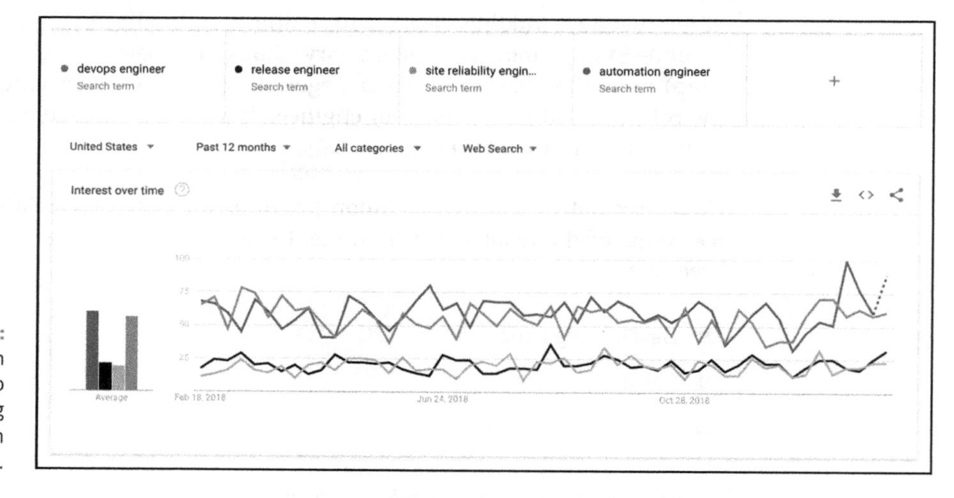

FIGURE 14-4: Common U.S. DevOps job titles adding Automation Engineer.

I wondered whether the popularity of Automation Engineer was unique to the United States, so I expanded my comparison to include data globally. Figure 14-5 shows the results of that comparison, and you can find the data for the figure at `https://g.co/trends/VDtFB`.

To substantiate the results from Google Trends, I looked at LinkedIn job postings in the U.S. At the time of writing, DevOps Engineer again came out as the clear winner, with Automation Engineer following shortly after. Here are some additional job titles and the corresponding number of job postings:

- DevOps Engineer: 4,918
- Automation Engineer: 3,316

- Site Reliability Engineer (SRE): 1,513
- Cloud Engineer: 1,403
- Infrastructure Engineer: 1,266
- Release Engineer: 610
- Sys Admin: 261

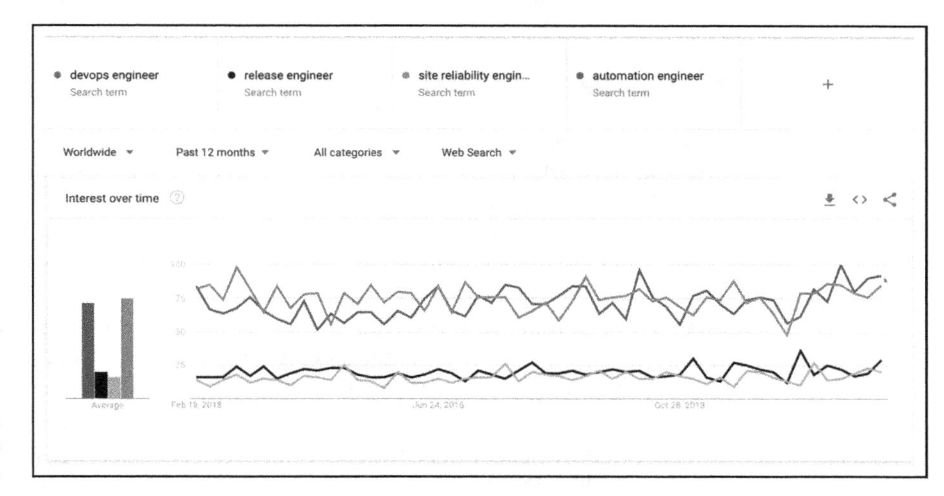

FIGURE 14-5: Common DevOps global job titles.

Recruiting Never Ends

Finding the best employees isn't as simple as putting an ad in the paper and collecting résumés, nor can you solve the problem by simply hiring a recruiter to do all the hard work for you.

WARNING

Recruiters can be a great asset during the hiring process. They are experts in interviewing, complying with employment laws, and negotiating a salary. However, because of aggressive hiring practices, many software engineers have grown to distrust recruiters. Reach out to candidates directly before introducing them to a recruiter. This simple step builds trust and shows that your recruiters are part of a well-thought-out hiring process. Without an introduction, you may lose candidates who worry that they're one of hundreds of candidates in a large pool that will go nowhere.

The key to hiring quality candidates is to always be . . . recruiting. That's right. This idea doesn't exactly sound like the pep talk you might remember from the movies, but it works. I keep two lists of people in my head at all times: people I would love to work with and people I would never work with. Each list grows as I meet more and more engineers.

Your lists are likely to be quite different from mine for a thousand different reasons, but start thinking about whom you would want to hire or work with if you ever got the chance. Then, when the opportunity presents itself, you'll be ready.

Finding the right folks

If your search for candidates is coming up short, consider your broader community and network. Think about the people you know directly and then also consider those second- and third-level connections — friends of friends who might be a good fit or know someone interested in the job. When you feel stuck, make lists of your personal connections and ask for community referrals.

The tech community is a thriving one. Well-attended conferences, meetups, and other get-togethers are happening every day. Reach out to organizers as well as influencers who are deeply involved in the community. They have enormous networks and are sure to know someone who might be looking. If not, ask whether they would be willing to post your job on social media or in their newsletter. (Many newsletters and podcasts need sponsors, which is a great way to get the word out about your position while supporting the community.) Also, don't forget to attend meetups and announce you're hiring. Meetups typically allow a few minutes at the start for job announcements.

TIP

Be pointed in your approach to ensuring that you create a diverse candidate pool. Referrals have a tendency to work against marginalized and underrepresented communities. Reach out to candidates who don't look like you, don't sound like you, and don't think like you. If you're not sure how to go about finding such candidates, ask for help. Hire one of the many consultants in tech who focuses on diversity and inclusion. Even a short engagement can help coach you and your team toward better hiring practices.

Word of mouth is a great way to find candidates. (It's how I've found every job I've had in tech.) Word of mouth networking is basic. In addition, I work hard to be as helpful as I possibly can because I currently have more reach than others. If you have connections in tech, use them to get people jobs. Few activities are more fulfilling and impactful.

Passing along great candidates

Sometimes you have to make a hard choice between two people who are interviewing with you. One of the best ways of keeping candidates in your solar system is to help those people find great employment. I can't hire everyone I want to hire because of resource constraints, but I love introducing them to friends, colleagues, and acquaintances whom I know are hiring.

For some reason, few people participate in this practice, but if you can make an introduction that leads to a great opportunity, you'll have made a loyal friend — one on whom you can count for referrals in the future.

TIP

Check with your company to ensure that engaging in this practice doesn't break some kind of internal policy.

Evaluating Technical Ability

The age of obtuse riddles and sweat-inducing whiteboard interviews is waning — and for good reason. If a whiteboard interview is facilitated by an engineer who cares more about tricking the candidate than they do about discussing a technical conversation, you'll go nowhere fast.

Whiteboarding interviews have taken a lot of heat recently for putting underrepresented and marginalized groups — which includes women and people of color — at a disadvantage. In this age, it's absolutely vital for tech companies to hire diverse workforces, so this situation is unacceptable. However, you have to somehow gauge a person's technical ability.

What's the answer? Well, the good news is you have options. (The bad news is . . . you have options.)

REMEMBER

How you hire will determine who you are.

Whiteboarding revisited

The whiteboard interview was never intended to be what it has become. In my first whiteboard interview, I was handed a computer program printed on eight sheets of paper. The instructions? "Debug the program." *Umm . . . excuse me?*

The whiteboard interview has become a situation in which you give a candidate a seemingly impossible problem, send them up to the board with a marker, and watch them sweat profusely while four or five people observe their panic. This type of interview provides no one with quality information on whether either the employer or the interviewee is a good fit for the other party.

Although others have called for the elimination of the whiteboard interview, I have a more nuanced suggestion: Change it. Make it a discussion between two people about a piece of code or a particular problem. Don't make the problem something crazy, such as balancing a binary search tree. Unless the job you are interviewing for is literally writing code in Assembly, you do not — I repeat, you do not — need to evaluate the candidate's ability to write Assembly.

Be cognizant of the job you are looking to fill, the skill sets required, and the best way to measure those skills in a candidate. Have a single engineer on your team sit down with the candidate and talk about the problem. How would you start the conversation? What problems do you run into along the way? How would you both adapt your solutions to the challenges you encounter?

This conversational approach accomplishes two things:

- **It reduces panic.** Most people don't think well under pressure. Plus, you don't do your job everyday while someone stares over your shoulder, criticizing every typo or mistake. You'd quit that job in an instant. So don't force people to interview that way. Instead, give your candidates the chance to show off what they can do. You'll gain insight into how they think and communicate.
- **It mimics real work.** The conversational interview gives you an idea of what it would be like to work with this person. You don't solve hard problems at work by watching each other struggle. (At least, you shouldn't. Really. That's not very collaborative or DevOps-y, leaving your colleagues to suffer in their silo.) Instead, you work together, trade ideas, think things through, make mistakes, recover, and find a solution — together.

The best whiteboard interviews are collaborative, communicative, and centered around curiosity — all the things I love most about DevOps.

Offering take-home tests

An alternative to a more traditional whiteboard interview is the take-home test. This type of test is particularly friendly to people who have any kind of anxiety or invisible disability that impacts their ability to participate in a whiteboard interview. This style of interview is also friendly to engineers who struggle intensely with imposter syndrome.

TIP

Imposter syndrome describes high-achieving individuals who struggle to internalize their successes and experience a persistent feeling of being exposed as a fraud.

A take-home test consists of some type of problem that a candidate can solve at home in their own time. Take-home tests are often set up as a test suite for which the candidate must write code to make the tests pass. Alternatively, the problem could be something relatively small, such as, "Create a program in [your language of choice] that takes an input and reverses the characters." The options are endless, and you can tailor the test to your tech stack as you see fit. You can even ask candidates to deploy their application. Ensure that you allow candidates to use open source tools or provide them with the necessary subscriptions to use particular technologies.

The major drawback to take-home tests is that you're asking people to take time during their evenings or weekends to do what is essentially free work. Even if you pay them for their work on the take-home test, this style of interview can unfairly impact someone who has other responsibilities outside of work, including caring for children, a partner, or ailing parents. Not every great engineer has unlimited time to commit to their craft. But if you limit your candidate pool to people who can afford to dedicate 5–10 hours to a take-home test, you'll quickly find your team becoming homogenous and stagnant.

Reviewing code

The interviews I love most are ones in which I sit down with an engineer, or a group of engineers, to solve real bugs in real code together. You can take a few approaches to a real-time code interview. You can mimic a take-home test and give the candidate an hour or so to create a program or write a function to make a series of tests pass. You can also stage the interview like a code review in which you pull up an actual PR and dig into what the code is doing as well as what could be improved. In many ways, the pair-programming nature of a code review combines the best parts of both a whiteboard interview and a take-home test — but without some of their major drawbacks.

TIP

Pair programming is an engineering practice in which two engineers sit down and work through a problem together. Typically, one person "drives" by owning the keyboard, but they collaboratively decide what approach is best, what code to add, and what to take away.

If the job position involves an operations-focused role, using this real-time coding approach is even better. Although many ops folks are learning to implement infrastructure as code or manage configurations, they don't have the same experience as developers. Reviewing what something does and how it might work is a fantastic way to confirm that the candidate has experience in the tools and technologies list on their résumé as well as ensure that the candidate can communicate with a team.

Firing Fast

Occasionally, someone who doesn't work out will slip through your interview process. They'll appear to be talented, collaborative, and an all-around great addition to your team. No one hires a candidate who is likely to be an undesirable employee. However, the results sometimes don't match the process.

You should act quickly in these situations. Allowing someone to exist on your team whom you know is not a good fit can have dire consequences. However, I also strongly believe in giving people a fair shake, which requires you to do several things:

- Explicitly communicate your expectations.
- Address deficits in performance quickly.
- Recognize and reward people who meet or exceed expectations.

If you're not a manager, you can still take these actions. Leaders don't always have a management title. You can absolutely have a conversation with someone about adjusting their work behavior, or thank someone for their hard work — no matter what your title is.

When you run into an issue with a new hire (or any employee or colleague), you should address the issue quickly and directly. Though interpersonal conflicts and poor employee performance can take many forms, three types of challenges arise the most frequently, as described in the following sections.

The jerk

No amount of brilliance can make up for an jerk's cost to the team. It doesn't matter if they're a genius, or if they created a tool, technology, or language. If they can't work with your team, they shouldn't be on it.

Jerks can destroy morale and drive great engineers away. In the worst-case scenarios, a manager's inaction to rectify a team issue serves to reinforce a team's belief that the situation won't be addressed. People will leave, often more quickly than you think.

Robert I. Sutton, author of *The No Asshole Rule: Building a Civilized Workplace and Surviving One That Isn't*, identifies two tests that you can use to recognize a jerk:

- Do people feel oppressed, humiliated, or otherwise worse about themselves after encountering the person?
- Does the person target people who are less powerful?

In his work, Sutton found specific behaviors that are often found in these workplaces. He termed this list the Dirty Dozen, which includes insults, unsolicited touching, threats, sarcasm, humiliation, shaming, interrupting, and snubbing. If you recognize any of this behavior on your team, take action as soon as possible.

The first step toward addressing a jerk is to clearly state that their behavior is unacceptable. Communicate your expectations for professionalism and mutual respect in all employees and explain what the consequences will be if the jerk doesn't adjust their behavior. Then, if you don't see improvements, consider removing them from the team (and possibly the company).

The martyr

This type of employee fights every disagreement to the death. The conflict could involve the language or framework to use to write an MVP, or whether to add a particular feature, or whether a bug arose from a code issue or from user error. The content of the disagreement doesn't matter.

Often, these martyrs either fear change or view themselves as a bit of a rebel. In their own way, they're trying to help guide the team toward the solution they feel is best. They don't behave this way out of malice. Instead, they have a communication issue. They have a tendency to talk over people or simply talk until other people give up.

The constant arguing and fighting that the martyr brings to a team is disruptive. It's a distraction and can be extremely time consuming — so much so that other employees may stop participating in conversations, especially in discussions around complicated problems or architecture.

The benefit of these employees is that they actually tell you what they think. This kind of problem is much easier to seek out and address than a problem caused by an employee who suffers silently. The trick to addressing the martyr is to be sensitive of their ego. If they resist team decisions passive-aggressively, call the behavior out in a private conversation.

The underperformer

Starting a new role always involves a learning curve, but occasionally someone's skills or efforts clearly don't match your expectations. This situation can be tricky. You want to avoid micromanagement and give employees enough room to succeed. However, you don't want to let someone's less-than-stellar performance become the norm.

Before acting, ensure that you've done the following:

- **Explicitly communicated your expectations about the role.** Be specific in your communication, as in "Complete *x* feature within *n* days," "Resolve *n* bugs each week," or, "Rearchitect our deployment pipeline using *x* tool."
- **Asked whether they need help.** Occasionally everyone gets in over their head. Sometimes people just need a nudge to get going in the right direction. Perhaps they're afraid of looking stupid. Starting the conversation without fear of repercussions opens the discussion and allows you to discover solutions together.
- **Ensured that the issue relates to performance.** During periods of my divorce, I was not the world's best employee. I had bad weeks. I felt emotionally empty and scared. You employ and work with humans, and humans get sick, or they have partners dealing with disabilities or chronic illnesses. They may have parents who get hospitalized, or have children whose caregivers bail at the last minute. Don't immediately conflate temporary poor performance with a chronic mismatch. Ask the employee, "Is there any stress outside of work that I can help with?" Sometimes people just need to vent and know they're safe at work.
- **Provided training.** Onboarding isn't easy, and I've been thrown into the fire more than once. If you're expecting your employees and colleagues to sink or swim, you're not doing your job. Explain the institutional knowledge you've acquired but have come to think of as second nature. Partner new employees with senior engineers who can serve as a single resource for questions. Be patient.

If you're confident that you've done your due diligence and are convinced that the employee's problem truly is a performance issue, communicate that fact. Request some time to sit down with them, one on one, and explain that they aren't meeting performance expectations. Be specific, as in "The code you've committed isn't meeting our quality standards because *x*." Or, "If you can't meet a deadline, I expect you to communicate the problem well before that date."

If an underperformer makes progress, reward it! That outcome is the best you could hope for. When a manager calls a meeting only to criticize performance but never recognizes good work, employees can quickly become demoralized. A simple "Thank you for your hard work" goes much further than you might think.

» Motivating engineers

» Creating self-motivating teams

Chapter 15
Empowering Engineers

Engineers are the engine of any tech business; they power the entire operation. Taking care of your engineers is vital to their continued health, happiness, and productivity. It behooves businesses to understand the motivations of engineers and create environments in which they can thrive. Happy engineers product better software, faster. It's just that simple.

This chapter focuses on scaling an engineering team through DevOps, motivating engineers to produce their best work, enabling your engineering team to allow engineers of diverse backgrounds to work collaboratively, and measuring your success.

Scaling Engineering Teams with DevOps

Growing your teams is one of the hardest challenges of tech — one DevOps attempts to assuage. I believe the greatest challenges in tech aren't technical, they're sociotechnical. Our systems have evolved beyond our machines. The challenges we face now have more to do with human behavior than with bits and bytes.

The type of business formed at the beginning stage of a startup is radically different from the one that evolves after years of trial and error — so different that you could say that each was a distinctive company. Scaling involves much more than simply adding personnel. You can't just make your startup bigger and announce you're an enterprise.

One of the challenges of scaling a company — at any stage — is communication. Figure 15-1 shows how quickly complexity in a system can grow. In the early stages, you're likely to be one of a few engineers, each of whom has a hand in building an application and its infrastructure. Perhaps you all work in the same room or are a merely quick call away from each other. The team and system are small and contained enough that you can keep track of all the moving parts in your head.

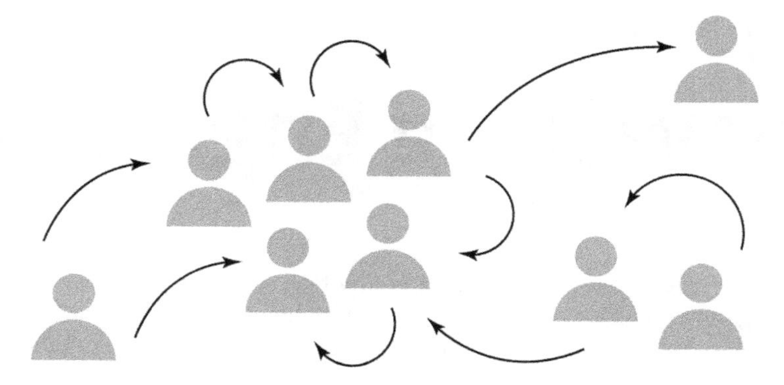

FIGURE 15-1: Communication complexity in large teams.

As you add people and components, more intersections of communication develop. A formula exists for measuring the total pathways of communication on a team: $n(n - 1) / 2$. For a team of 5 people, there are 10 communication pathways. This size is entirely manageable. It escalates quickly, though. A team of 100 people has 4,950 communication connections, which is overwhelming.

Scaling will almost always be a messy process, and there's no way around that likelihood. But this chapter looks at some organizations outside of tech that managed these kinds of growing pains admirably, and how you can do the same through DevOps.

Three stages of a company

I give a talk called "Scaling Sparta: Military Lessons for Scaling a Development Team," in which I compare the three stages of a company to three militaries: the Spartans, the Mongols, and the Romans. Spartans serve as my analog for a startup. Evaluating the Mongols can illustrate what it means to thrive as a late-stage startup or mid-sized company. And, finally, Rome serves as an ancient example of an enterprise.

Scaling is an important facet of DevOps because companies aren't stagnant. You experience periods of high growth as well as contractions. You'll likely see your company hire, and fire, over the years. Staying true to your DevOps principles and continuing to evolve your processes will define your success through those stages of expansion and contraction.

Startup

At its largest, the Spartan army numbered around 10,000. Despite its modest size, Sparta was obsessed with war and nurtured this obsession in a child from the moment that child was born. Throughout their lives, Spartan citizens had to prove their worth to their nation and their unit. In many ways, their lives were defined by their usefulness on the battlefield.

The Battle of Thermopylae pitted 300 Spartans against (supposedly) 100,000 Persians. In one sense, that's the challenge that a startup faces. How do you iterate quickly to challenge companies with 10 times your resources? Here's how:

- **Differentiate yourselves.** Identify the product or value-add you do best and do that really well. Adding pointless features exhausts your engineering team and frustrates your customers. More rarely equates to better.
- **Hire generalists.** At the startup stage, you need your engineers to have pretty good and diverse skills. Look for engineers with a natural curiosity and a willingness to dive into new technologies. You want people who can adapt and solve problems using any tool — not necessarily the one they're an expert in.
- **Be bold.** Do the scary thing. Take risks. Disrupt. Be inventive. The advantage of being small is that you're agile; you can adapt more quickly than your more established competitors. Use that agility to your advantage and stay small for as long as you can.

Late-stage startup or mid-sized company

The Mongols are a fascinating military to study, and their success is forever credited to Genghis Khan. To this day, it remains the largest geographic empire in history. At its peak, the Mongol army was 100,000 strong. They conquered more land in 25 years than Rome did in 400. The brutal conditions in which the Mongols lived made them tough and resilient — the very qualities that late-stage startups need to survive through the awkward teenage years.

Under Genghis Khan, the Mongols embraced religious tolerance. In other words, they accepted people as they were, allowed them to thrive, and respected their autonomy. They promoted people on merit. If someone did well and benefitted the greater tribe, they reaped rewards. Mongol society was egalitarian. Both men and women contributed and everyone's work was respected. Finally, Mongols adapted quickly. Living on the land in rugged plains, they had no experience with walled cities but mastered the art of siege warfare.

Genghis clearly wasn't without flaws, and I don't recommend following in his homicidal tendencies, to say the least. But his leadership has some key qualities from which to glean important lessons:

- **Focus on the what.** Not the why. I've said this before and I'll say it again. It is vital that you set the mission for your company, your organization, and your team. Wherever you sit on the chain of command, set the mission — even if that mission is for you and you alone.
- **Enable autonomy.** Allow everyone the room to complete the mission as they see fit. Mongol soldiers kept three to four horses at all times, which permitted them to travel long distances at fast speeds without exhausting their animals. Give your engineers the tools and resources they need to thrive.
- **Think strategically.** Reactionary businesses rarely survive. Colonel John Boyd of the U.S. Air Force created the OODA loop, which stands for observe, orient, decide, and act. Use this method to think through problems and prevent yourself from being slowed down by fear, uncertainty, and doubt.
- **Keep structure simple.** If you're a late-stage startup or midsized company, you need to introduce structure but keep it simple. Start with a quick daily standup. Track features and bugs using a simple tracking tool. Avoid as much complexity as possible, especially if the complexity exists only to make you feel more like a "real" company — whatever that means to you.

Enterprise

Rome is perhaps the greatest example of an enterprise organization ever known to humankind. In a large-scale organization, reliability and predictability outweigh the novelty of taking risks in hopes of big rewards. In an empire, protecting what you've built becomes increasingly important.

The activities most abhorred by startups — administration, management, and process — are the bread and butter of enterprises, and for good reason: The potential pathways for miscommunication at a company of 100,000 employees is astronomical. (I'll spare you the math; it comes out to 4,999,950,000 links between people. *Whew.*)

Rome was an incredibly complex civilization, and compressing its key military strategy into a section of a book is impossible, so I focus on the Imperial Roman army under Caesar Augustus, which is when the Roman army reached its peak size of nearly half a million soldiers. Rome started by giving its soldiers grants of land. As those resources became more scarce, they switched to awarding soldiers with a set amount of denarii (roughly 13 years' salary) after their service. Sounds a lot like a pension, doesn't it? They divided the military into three components:

- **Legions:** Heavy infantry made up of Roman citizens. Soldiers served terms of 25 years and conscription was used only in emergencies.
- **Auxilia:** Troops recruited from noncitizen residents called the *peregrine.* These soldiers held positions in infantry, cavalry, archery, and special forces. At the end of their service, auxiliaries were awarded Roman citizenship.
- **Numeri:** Mercenaries from allied tribes outside the Roman empire. I like to think of these folks as contractors.

Organizationally, Rome split its command into provinces, overseen by legion commanders called *legati* who reported to the provincial governor and then up the chain directly to the emperor in Rome.

Legions had a higher status but depended heavily on the auxilia for support on the battlefield. This dependence reveals the importance of methodologies like DevOps in large organizations. DevOps doesn't promote the idea that everyone needs to do every job. Instead, you need to have a general understanding of every job and — here's the really important part — respect the people who do the jobs you don't.

The lessons Rome teaches are endless. (If you love history, as I do, I highly recommend that you look further into each of these fascinating militaries.) Meanwhile, here are the key takeaways from the Roman enterprise for this book's purposes:

- **Break into small teams.** Think about your application. When a component becomes too large, you break the logic into smaller pieces. The same principle applies to the people on your teams. Small teams enable engineers to move quickly and effectively, which makes intuitive sense for a DevOps organization. Sharing and collaboration — key principles of DevOps — become impossible past a certain scale. You must provide your team with the framework to make these rather lofty goals possible.
- **Allow independence.** Each unit in the Roman army had its own standard — represented by a pole with a variety of decorations. The practical use was to visually communicate where the bulk of the unit was located on a vast battlefield, but the standards held deep meaning for the soldiers. They believed that their standards represented a divine spirit, and they prayed to it.

Each unit also had its own unique culture. Applying that "small" approach to your own teams will permit like-minded people to work seamlessly together.

- **Master logistics.** Rome invested heavily in an extensive and well-maintained road system. This infrastructure allowed the transport of troops and supplies throughout the vast empire. This approach is perhaps best summarized by Facebook's latest internal motto: "Move fast with stable infrastructure." The infrastructure of Rome was one of its greatest advantages over its peers.
- **Invest in your employees.** The longer your employees stay at your organization, the more informed every team becomes on institutional knowledge and systems. Despite all attempts to get engineers to document everything — and you should encourage this practice — people naturally internalize information. This type of knowledge is the most valuable because people don't even think about it as knowledge. To retain employees, encourage work-life balance and pay them fair salaries. Avoid burning them out at all costs because it will cost you heavily in lost productivity.
- **Introduce specialists.** At the enterprise scale, specialists become an advantage and should have their place alongside the generalists that you hired during earlier stages of your company. In addition, allow generalists to become specialists through training and employment opportunities. Get to know the members of your small teams and understand their goals, both professionally and personally. Keep those goals in mind when making changes, providing new career opportunities and thinking about continuing education.

Motivating Engineers

Startups have a hard time transitioning into a large company, and that's because it's really hard. The quick wins you experience daily, sometimes hourly, at a tiny startup fade as you grow. Your plans extend from surviving one hectic period of time to thriving over weeks, months, and even years. Features become more complex and therefore slower to release.

This transition from the quick pace of a startup to the slow churnings of an enterprise can be demotivating to everyone. You simply don't get the same dopamine rewards at large organizations that you do at small, scrappy companies. If you're not careful, this lessening of rewards can cause motivation and overall job satisfaction to plummet.

Daniel Pink has compiled some of the best research on this topic in his book, *Drive: The Surprising Truth about What Motivates Us*. In many ways, science has proved the business methods of the 20th century to be simply wrong and often counterproductive. You don't work in a factory and you don't need a line boss. You work in an intensely intellectual and creative industry within a greater knowledge economy.

Researching motivation

Whatever your size, I suspect you want to use DevOps to become more productive, as well as to attract the best talent and outwit your competitors. To accomplish those goals, doing things the same way as they've always been done isn't going to cut it. In *Drive,* mentioned previously, Daniel Pink focuses on three principles:

- Autonomy
- Mastery
- Purpose

Simply throwing money at your employees isn't sufficient (although that fact won't stop some companies from trying). Instead, you must delve into their motivations.

Some of the greatest challenges in a DevOps transition involve motivating your team to produce high-quality work quickly and changing their unhelpful or harmful approaches to problem solving. The best managers aren't people who tell their employees what to do and then enforce that with a carrot or a stick. Instead, they persuade their employees to self-motivate. They encourage their employees to think independently and to get excited about their work.

You need to be aware of the two categories of motivation: intrinsic and extrinsic. Intrinsic motivation is the kind that pushes people to take action based on internal drivers. People naturally do work that they find rewarding. Extrinsic motivation is work done to obtain a reward or avoid a punishment. The former, intrinsic, is a much more powerful and long-lasting form of motivation.

If you view your employees as naturally lazy, I encourage you to step away from management. Perhaps no attitude is more damaging to the delicate nature of human relationships than contempt. Engineers don't sit still on weekends. Yes, they watch Netflix; I do, too. But they also take up sports, commit to open source projects, play with their kids, race cars, cook, and any number of other energy-demanding activities. Engineers aren't great at sitting still. They're thinkers and tinkerers. Some of the most knowledgeable people I know outside of tech are engineers. Never underestimate an engineer's ability to dive deep — I mean really deep — into a new hobby. They read every book about a topic that interests them, unlock every secret, experiment, break it apart, and become an expert.

Much of the research that Pink highlights originates with Mihaly Csikszentmihalyi, summarized in the book *Flow: The Psychology of Optimal Experience*. Csikszentmihalyi found that people enjoy the feeling of pursuing a difficult endeavor and attempting to accomplish a task they believe to be worthwhile. Through an incredible amount of research, he found that people thrived in the experience of pursuit and purpose — a feeling he described as flow.

DevOpsing motivation

You can use the DevOps approach to put your engineers into an environment in which they can achieve the flow mentioned in the previous section. Daniel Pink distilled Csikszentmihalyi's research by suggesting that managers and employees seek out "Goldilocks tasks." Such jobs are neither too easy nor too hard; they're just right. Finding this tension between extreme stress and extreme boredom is a tricky but worthwhile pursuit in your workplace, and it's one that the DevOps approach supports. By permitting your engineers the autonomy to own their work and feel pride in their contributions to their team, you enable them to achieve purpose.

Engineers — and all humans — want to be the masters of their own lives. They want to feel as though they are making decisions and have a reasonable amount of control over their lives and their work. This is autonomy. In addition, they thrive when they can continuously improve throughout their career. Continuously improving is a bedrock principle of DevOps. Engineers want to get better at what they do, and then keep getting better and better. This is mastery. Finally, engineers desire purpose. They want to know that their work and their contributions have meaning beyond the basics of subsistence. This is purpose.

Fully consider these three principles of motivation: autonomy, mastery, and purpose. Think about the last time you felt truly fulfilled. When you think about that time, what were you doing? Who were you doing it with? Did these principles play a role in your happiness?

Avoiding reliance on extrinsic rewards

You can think of a reward as the dangling of the carrot; it's what people give others when they do a job well. Rewards come in many forms: money, public recognition, and other awards. The challenge of rewards is that they change the way people's minds work. Dan Pink highlights a study he calls the "Candle Problem," a social experiment in which participants are asked to attach a candle to a wall with a box of thumbtacks and matches. When participants were offered money for performance, they solved the problem more than three minutes slower than the control group that was not offered a reward. You can watch Pink explain this phenomenon in his TED talk at https://www.youtube.com/watch?v=rrkrvAUbU9Y.

This result of offering rewards runs counter to what people have been taught about business. The more you pay employees, the better they perform, right? Actually, no. You can apply this psychological research to tech and understand it as one of the key values of DevOps in your organization.

REMEMBER

Offering a fair market wage and benefits is a baseline for ensuring that your employees can support themselves (and their families). But money is not a reward. Pay your employees what is fair. By providing a fair salary, health insurance, retirement plans, and other benefits, you remove a key stress in people's lives: money. No one wants to think about how they're going to pay a bill, buy a new car, put their kids through college, pay for their parent's cancer treatment, or survive a divorce. Your goal is to both remove the stress of having too little money while also eliminating money as the only reward for good work.

Autonomy

DevOps flips the switch on traditional management. Instead of deciding on a detailed course of action and then instructing your employees to do the work, DevOps managers create a vision and allow their employees the autonomy to create a plan for the work. Engineers work together to plan features as a team, think through the potential pitfalls and concerns with engineers from other areas of expertise, and then divide the work in a way that suits their strengths as a team. This approach is a powerful way to give your engineers autonomy. You allow them to direct their own work, and that empowerment pays dividends.

Mastery

DevOps creates an environment of rapid iteration and continuous improvement. You can divide this continuous improvement into five categories:

- Continuous planning
- Continuous development
- Continuous delivery
- Continuous feedback
- Continuous learning

You can probably add even more, but that forward movement is what satisfies the basic human need for mastery. Your engineers are empowered to take control over their work and hone their skills. No one wants to feel stagnant, and I can't think of something more stifling than being given a list of features to implement, handing them off, and then grabbing the next assignment. This is the traditional tech environment that DevOps is changing.

In that environment, engineers are little more than code monkeys. They write code, deploy it, and go home. Ugh. It makes me sad just thinking about it. Most engineers have worked (or known someone who worked) at a company that believed in this process. In such an environment, engineers are powerless to take pride in their work, appreciate the bigger picture and mission, and take ownership over their process, which undermines their ability to grow. Engineers in such a situation will either leave or stagnate — neither of which is a desirable outcome.

Purpose

You may think of having a purpose as meaning to work for a great cause. Don't limit your understanding of purpose to writing software for charities or working for a nonprofit whose mission is to solve hunger, though. If that's your passion, great! Do that. But you can find purpose in more ways than are obvious. Giving engineers time to mentor less experienced engineers on the team or in their community is one way of giving them purpose. Allowing them time to work on open source software while at work is another. You can also give people time to prepare and give talks at conferences or run meetups as well as hire people who care about your customers. However you enable it, a sense of purpose keeps engineers working hard — even when the project is hard and the challenges are overwhelming.

Monetary rewards seem to offer an easy fix to the problem of motivation. They're not effective, however, and can potentially have negative impact on your team's productivity and motivation. Your job — whether you're a manager or an individual contributor — is to look beyond the easy fixes and find what works. The individuals will vary as much as their preferences, but at their core, all engineers desire independence and purpose. Protect them from the elements of your organization that would deny them that fulfillment and you'll find yourself with a happy, productive team.

REMEMBER

People don't quit jobs. They quit managers.

Making work fun

The engineering industry constantly measures and compares people based on their technical skill and knowledge. In a healthy environment, this atmosphere can serve as a great way to keep people sharp. Grindstones keep blades sharp. But when an environment takes a turn from healthy collaboration to competition, the fun of learning is stripped away and replaced with fear.

Engineers who fear looking stupid are less likely to pick up new skills, attempt a new language, or suggest a new tool. New technologies take them out of their comfort zone. Their productivity slows down and initially they struggle. You must accept an initial slowdown when encouraging engineers to continuously learn and improve. Speed, after all, is not the only indicator of a healthy and productive engineering organization, and curious engineers can thrive only in healthy, fun environments.

Allowing people to choose their teams

At age 20, Spartan soldiers became eligible to join a *syssitia*, a sort of club. Members of the syssitia had to vote to accept a man into their group and the vote had to be unanimous. The Spartans may have been onto something with this approach. Allowing engineers choice as to whom they work with and what they work on can be a powerful tool in your quest to increase autonomy.

WARNING

Although I believe in the fundamental quality of allowing people to work with whomever they prefer, an opportunity for practicing exclusion arises with this approach. If you try this route, watch carefully to ensure that people aren't bifurcating along lines of social diversity — race, gender, ethnicity, religion, and sexual orientation. This behavior is a warning of larger challenges existing under the surface, and you need to step in as soon as you see it.

Bonding with the people you work with is critical, and not just on a professional level. You want to be able to like, admire, and respect them as people. Healthy teams don't know each other's birthday because it's their job but because they care about their colleagues. The same goes for knowing each other's kids' names, hobbies, and personal stresses. When you allow engineers the ability to choose their teams, you enable a deeper level of bonding through self-selection.

Measuring Motivation

As with everything in DevOps, tracking your experiments and measuring the output is critical to continuous improvement. I highly recommend that you survey your team regularly to ascertain their general level of happiness, motivation, and job satisfaction. As you measure your progress in the happiness department, track productivity during the same time. If you don't add productivity as a part of your data, it can be easily dismissed by others. Not all executives are created equal, and some still thrive in the old-school way of thinking about business and motivation.

Ensure that your way of measuring productivity is aligned between teams. Measuring developers on features shipped and operations folks on flawless deploys is not a DevOps-like approach. If measured simultaneously, these goals can quickly become a source of friction. Instead, use a measure such as the number of user stories (descriptions of features from the user perspective) released to customers. Focusing on user stories that make it to production emphasizes delivery as a team and removes siloed responsibility and incentives. My guess is you'll see a substantial increase in your team's overall productivity.

4 Practicing Kaizen, the Art of Continuous Improvement

IN THIS PART . . .

Improve your on-call procedures, manage incidents better, and minimize processes that lead to human error.

Prepare your systems to fail well, and embrace a growth mindset by learning from failure through productive post-incident reviews.

Consider the contributing factors of failure and how to run a post-incident review.

IN THIS CHAPTER

» Failing fast (and well!)

» Embracing a growth mindset

» Reviewing incidents as a team

Chapter 16

Embracing Failure Successfully

One of the gifts of software engineering is that the industry has emerged far later than other engineering disciplines. If you look to those older, more experienced industries, you can see many of the problems you face have been solved — or at least identified. (And isn't it nice to put a name to a problem?)

At some point along the way, executives adopted this concept of fail-fast and tailored it to startups. (You may recognize the term from Eric Ries's book, *The Lean Startup*.) Although perhaps overused and misunderstood, failing fast originates from system design. A fail-fast system quickly notifies the administrator of any indication of failure. This requires advanced detection of even a whiff of danger. These systems verify state along the entire process to ensure safety.

In this chapter, I dig into the origins (and misconceptions) of the commonly heard phrase *failing fast*. Chapter 17 offers ways to prepare for failure and learn from mistakes, and Chapter 18 tackles post-incident reviews.

Failing Fast in Tech

In software, a fail-fast system is ideal. In modern development, system components act independently and can change behavior if a failure is detected in a neighboring component. These features can make your system more fault tolerant, allowing it to function even as failure is occurring.

If you implement your system well with failure checks at each potential breaking point, it will show failure earlier than would be typical because you're made aware of the failure far before a cascading series of failures can cause catastrophic consequences. In other words, each component is treated independently in failure detection, so a domino effect is less likely to occur.

Failure checks provide more information about the issue, and closer to the source of failure. How many times have you triaged an outage or thought you had fixed a bug and assumed that everything was fine, only to discover — usually hours later — that the issue was caused by another component in the system, sometimes completely unrelated? These service interruptions are costly, so determining where a bug or outage originates from pays dividends well beyond the initial cost of architecting a fail-fast system.

WARNING

When I say "source of failure," I do not mean root cause. Complex systems simply have no root cause. They have only triggers of failure — that is, the final steps in cascading errors. Executives typically love (or demand) to know a root cause because it's simpler to take to the board and customers as an explanation. It's up to you to explain why attempting to determine a root cause is a flawed exercise. Read more in the section, "Going beyond root cause analysis," in Chapter 18.

Failing safely

A step beyond fail-fast is *fail-safe,* which is a system that shuts down operation immediately on discovering a failure to ensure the safety of humans, equipment, data, and any other assets that could be damaged. For example, had Knight Capital implemented appropriate checks on trading, the system would have halted operations and prevented the catastrophic failure of losing $440 million in under an hour. (See Chapter 12 for more details on the Knight Capital financial catastrophe.)

Creating a fail-fast system is less complicated than you might think. It simply involves thinking about handling failure rather than attempting to avoid it at all costs. In software, a fail-fast component will fail at the first sign of a problem. It could happen when a user inputs bad data into a form, for example. Rather than fail at the database layer (or later!), you design the form to ensure data quality before transferring the data to other components. Code designed to fail fast is easier to debug, reduces the number of components involved in a failing process, and prevents lag before the user receives an error message.

REMEMBER

The opposite of fail-safe is fail-deadly. Although you probably don't engineer software for ballistic missile submarines, nuclear reactors, or pacemakers, juxtaposition of fail-safe and fail-deadly is worth considering. Your result may not qualify as fail-deadly but instead be fail-fail or fail-bad or — my personal favorite — fail-fired. The decisions you make have repercussions, so considering the knowns and unknowns of your system is a valuable exercise.

Containing failure

Much of the preparation for failure isn't about trying to avoid failure. Instead, the idea is to expect and control for it. Though exceedingly rare in software systems, catastrophic failure occurs in systems that fail badly because of a single point of failure. This one Achilles' heel, if shattered, brings down the entire system. An example of a catastrophic failure is the 1836 fire at the U.S. Patent Office. The system had no data redundancy, and any patents lost in the fire were lost forever. (Fittingly, the patent for the fire hydrant was destroyed in the fire.) A more recent example of failing poorly, this time in civil engineering, was the Nipigon River Bridge. As a result of a partial failure, the bridge outage completely disconnected road access between eastern and western Canada. No alternate route existed along the Trans-Canada Highway.

Ship hulls are containerized (pun intended) to allow for a breach of water in one without sinking the entire vessel. Elevator brakes are fail-safe because tension from the elevator cable above the car holds the brakes away from the brake pads. If something severs the cable, the brakes latch and the elevator comes to a stop.

THE OTIS SAFETY ELEVATOR

Elisha Otis first showcased his elevator brakes in 1853 at America's first World's Fair, held in New York City. He rode an elevator platform high above the cheering crowd and ordered the rope that held him be cut. You can imagine the excitement and tension building in the crowd as people watched this demonstration unfold. I like to think of a collective gasp rippling out as the rope was severed and Otis fell, for a moment, and then halted to a complete stop (`https://www.6sqft.com/elisha-otis-now-162-year-old-invention-made-skyscrapers-practical/`).

You can consider data centers to be fail-safe. Cloud providers achieve 99.999 percent availability by implementing *n*+2 redundancy. Data is stored in three separate places to allow for access even in the event of failure in one server while another is down for planned maintenance. That type of redundancy is expensive but worth it for some companies.

Accepting human error (and keeping it blameless)

Human error is a flawed term. It implies that humans can be the single source of failure of an incident, rather than the complex sociotechnical systems in which those humans operate. If something goes wrong, and you determine a human was the "root cause," you just fire them, right? Problem solved!

Nope. Wrong. You cannot fire your way to a great engineering team. If a human was allowed to err, your system failed, and you have a system problem, not a human problem. Humans are catalysts that exist within a system. Although humans make mistakes, they are part of a whole. One of your goals in a DevOps culture is to eliminate the possibility of human error. Although planning for every potential failure is an impossible task, chasing the goal will improve your systems and processes immensely.

Failing Well

The most important aspect of a DevOps-focused engineering team is the ability to fail well. This ability has more to do with the people than with your tooling.

I talk at length in this book about iteration and continuous improvement. The Japanese word *kaizen* means *improvement* or *change for the better*. In the context of DevOps, kaizen means to continuously improve all areas of your business. This ancient concept of kaizen is applied in lean manufacturing and The Toyota Way.

Kaizen isn't some grand, sexy process. Instead, it refers to tiny decisions, made every day, to slowly improve productivity and eliminate wasteful work. Lao Tzu captured this beautifully in the *Tao Te Ching*: "The journey of a thousand miles begins with one small step." This work to improve daily can't succeed by one individual's efforts alone. Like any DevOps transformation, it requires adoption by the entire team. Everyone in your organization must pitch in and take ownership of the process by applying the philosophy of kaizen.

The most interesting aspect of kaizen is that it embraces failure. Not catastrophic failure, mind you, but it accepts that the process isn't perfect and you always have room to refine and improve. The realization that everyone has a role to play in continuous improvement is a healthy first step in establishing accountability as a team. People usually think of accountability punitively, but in DevOps, accountability means to take ownership over your work, your team, and your organization. Everyone, from the most junior engineer to the CEO, has the ability to have impact. Using the kaizen approach, everyone makes small changes, monitors results, and adjusts continually.

Maintaining a growth mindset

In her book *Growth Mindset,* Carol Dweck describes two groups of people: those with a fixed mindset and those with a growth mindset. Dweck stumbled on this finding when conducting research on students' response to failure. Dweck and her colleagues observed that some students rebounded from failure and others were crushed by the weight of it. When they dug into the underlying belief structure that resulted in these outcomes, they realized that some students viewed failure as a necessary step in learning whereas others felt that failure was indicative of whether they were good at something. In the former, failure is a stepping stone. In the latter, failure is a definitive and terminal blow. Figure 16-1 steps through the process that someone with a fixed mindset goes through when confronted with failure. Contrast Figure 16-1 with Figure 16-2, which shows the internal dialogue of someone with a growth mindset.

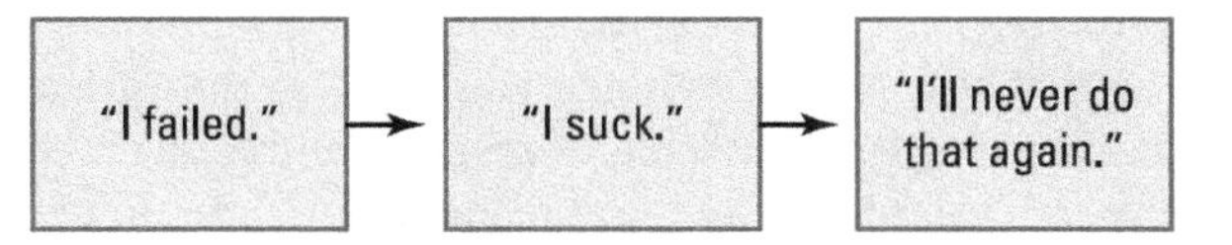

FIGURE 16-1: The thoughts of a fixed mindset.

A fixed mindset isn't an immutable state. I started life with a fixed mindset. I struggled initially with math, especially when under time constraints. Although studies involving history, people, and language came easily to me, math never did. It still doesn't. My brain has to fight to understand the concepts and solve the problems. I naturally concluded that I simply wasn't good at math and shouldn't pursue it. Any hesitation I may have experienced was silenced when my trigonometry teacher called me "stupid." I proceeded through the next 10 years of my life thinking I was terrible at math and should never be allowed anywhere near math, science, or technology.

It wasn't until I started at code school that I fully accepted a growth mindset. I was desperate to finish and change careers, and that provided me enough motivation to push through the rough patches. Then something remarkable happened.

I learned. I'm still not a math savant. I can't add two numbers together under pressure. I simply freeze. But guess what? I'm a great engineer. And when I need to learn a math concept to do my job, I do just that. I learn. No matter how uncomfortable the process feels.

You can learn to develop a growth mindset as you can any other skill. It simply takes time to adjust your thinking from reactionary and defeatist to curious and optimistic.

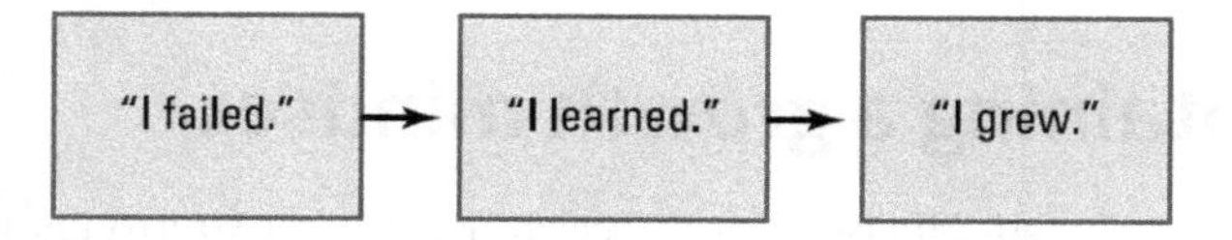

FIGURE 16-2: The thoughts of a growth mindset.

Creating the freedom to fail

Most of the time when you try something new, you fail, and there's no shame in it. The first time I tried to drive a stick shift, my mother's grey truck lurched forward and stalled in the street in front of our driveway. I didn't even bother to try to park it. I simply got out of the car, returned to the house, and left my father to deal with the car left in the middle of the road. It took me another eight years to buy a car with a manual transmission and finally learn. (My mother had to drive it off the lot for me. I'm stubborn.) But now I have two six-speeds and can't imagine driving an automatic.

The first software program I ever wrote was a 200-line mess of nested loops. It was so bad that it would have made you cry. But then I learned OOP (object-oriented programming) and division of responsibility. The code I've written over subsequent years is significantly more readable and functional.

You likely have countless stories like mine: moments of temporary setback, failure, and then a (sometimes) arduous process of slowly learning. But unless you cultivate a company culture that embraces failure, your engineers will continue to live out the patterns they've learned throughout their lives:

- Reduce risk as much as possible.
- Avoid failure at all costs.
- Cover up mistakes.
- Look to blame others instead of accepting accountability.

Everyone carries the scars formed from a lifetime of bad bosses, stressful jobs, and personal hurt. Part of being a great manager and colleague is recognizing those scars and working to create a safe work environment while respecting the fears and knee-jerk reactions of those around you.

Encouraging experimentation

When left alone, and given the time and resources to experiment, engineers make all sorts of fascinating discoveries. Google's famous 20 percent time — which gives engineers one day a week to work on whatever project they want — resulted in Gmail and AdSense, two significant products.

Atlassian takes the concept of 20 percent time a step further and allows different teams to implement their own versions of innovation time. Some teams have an innovation week once every five weeks. Also, the entire company can participate in a 24-hour hackathon called ShipIt during which they create and deploy both technical and nontechnical projects in a day.

However you choose to implement it, encouraging experimentation is core to creating a safe work environment that embraces failure.

Balancing challenging work with fulfilling achievements

In his book *Drive*, Daniel Pink describes the research of Mihaly Csikszentmihalyi, who found that people who were guided by purpose experienced a feeling he called "flow." This feeling requires what Pink calls "Goldilocks tasks" — work that is neither too easy nor too hard, but just right. When you experience flow, you are fully immersed in your work, driven forward not by your salary or potential extrinsic rewards but instead by the purpose you derive from your work.

If you're an individual contributor, seek out tasks that you find challenging but that don't throw you into a spiral of stress-induced self-doubt. Finding this balance isn't easy and involves trial and error. The trick is to trust yourself and your colleagues enough to ask for help when you need it and to take on increasingly harder work when you're ready.

If you're a manager, note that helping your team achieve balance is how great managers overtake those who care more about this quarter's bottom line than the longevity of their team. Talk to your engineers. Find out where they feel confident and where they feel they need some work. Verify these qualities in your team members yourself or ask your senior engineers to assess their colleague's strengths and weaknesses. The goal here isn't to root out the runt of the litter. Instead, use this activity to balance the team as a whole. Everyone has both strengths and weaknesses as an engineer and an employee. Being honest about those and structuring the team for balance and growth is essential to reap the benefits of DevOps.

Rewarding smart risk-taking

Failure is an outcome. Risk-taking is an action — the input in an innovative culture. Discussing smart risk-taking and setting boundaries is a healthy part of planning for failure. Managers can model this behavior by taking small risks themselves. When managers embrace innovation and shed the instinct to maintain the status quo, they create a culture of taking chances and learning from failure.

Part of smart risk-taking is controlling the blast zone, which is the radius of services or users affected by a failure. Instead of allowing every user access to a test in production, release it to a small set of randomly selected customers. Another aspect of smart risk-taking is thinking through the potential wins and losses from a particular experiment. Release small changes frequently for the best results.

The late Randy Pausch of Carnegie Mellon University often reminded students that one penguin had to be the first to make the plunge into the water without knowing what predator might lurk just beneath the surface. He rewarded the "First Penguin" award to the student who took the boldest risk during the semester. I love this and wish every engineering team had a reward like this because it makes failure light-hearted and normal. It makes it part of everyday life instead of something foreboding and to be avoided at all costs.

Building a soft landing

You can't kick the baby bird out of the nest and then leave it to struggle on its own when it doesn't fly on the first go. You must shield your team from the consequences of failing from stakeholders outside engineering. This includes executives, colleagues from other departments, and sometimes customers.

REMEMBER

If you're a manager, your job is to provide cover for your team to do their work. Be confident in the larger, long-term benefits of having a culture in which failure isn't something to be ashamed of. A learning culture pays dividends in employee job satisfaction, innovation, and collaboration. A team that successfully fails together has a level of trust most teams will never achieve.

Perfecting the art of done

You can rewrite bad code, but you can't rewrite *nothing*. Perfectionism is the greatest enemy of productivity because of its paralyzing effect on work. You become so afraid of being bad at something that simply doing nothing at all feels better.

Logically, you know that doing nothing isn't wise. Your neocortex recognizes that you need to do the work. Yet, the oldest part of your brain fights against doing it. The oldest part of the human brain skews heavily toward safety, and doing nothing is safer than doing something and risking failure.

Getting a first draft done is much more important than making that draft even remotely high in quality. If you're a developer, you know all too well how you struggle with a hard-to-solve problem. You write a hundred lines of embarrassingly poor code. It mostly works. Yay! Now that you know *how* to solve the problem, you can throw your first draft away and rewrite the solution in a few lines of divine code.

Give yourself (and your team) permission to suck. It's not a permanent state but instead is part of a larger journey. The faster you allow yourself to suck, the faster you get to work and, ultimately, perfect your art of getting work done.

Failure is wasteful only if you don't learn from it. A learning culture goes hand in hand with an engineering team that accepts and encourages taking risk that may result in failure.

IN THIS CHAPTER

» Minimizing the processes that lead to human error

» Improving on-call response

» Managing incidents when they occur

» Measuring your success

Chapter 17

Preparing for Incidents

What's an incident or service outage? Good question! Essentially, an *incident* is any technical disruption of your business. Incidents come in all shapes, sizes, and severities. For example, if your business is banking, members of your financial institution might not be able to access their bank accounts online. If your business is online photo storage, a potential incident might prevent users from uploading new photos. If your business is retail, maybe users can't make purchases because your payment processor is down.

Sometimes an incident can be rather tame. Perhaps the "Add to Cart" button is duplicating requests and adding two items to customers' carts instead of one. Irritating, yes. But the situation isn't dire because the customer can edit the quantity in the cart. Other times, incidents can be much more traumatic. Perhaps your sign-up form is preventing users from joining your site or your payment processor service is down. Or a database error has erased critical user information. Yikes!

In this chapter, I show you how to prepare for incidents and service outages of all kinds. I walk you through how to ensure that your processes reduce the possibility that humans will be the cause of an incident, how to better prepare your on-call team for incident response, and what to do when an incident strikes. Along the way, I share a few use cases that might help you better prepare yourself and your team for when an inevitable incident hits.

Combating "Human Error" with Automation

Human error can quickly lead you to believe that humans are the "root cause" of a failure. (Read more in Chapter 18 about why root cause is a problematic term.) Instead, if a human happened to be a trigger of failure, look at the situation this way: Judgments and decisions made by an engineer may have contributed to the disruption. Perhaps even more important, consider that the systems and processes of your engineering organization led to (or did not prevent) those judgments or decisions.

Incidents will always be a part of developing and maintaining software. People are only human. It happens. Stuff breaks. The problem with incidents isn't that they happen. Yes, this reality is unfortunate and uncomfortable, but the real issue is that the same incident (or incidents that are eerily similar) consistently reoccur. These incidents are often long, drawn-out, and stressful events for everyone involved — including customers — and they often repeat themselves.

By now you've likely realized that more often than not, humans are the challenge in DevOps. "Human error" is the label that people put on the common (and frequent) occurrence of human mistakes. If you're thinking that the solution for incidents that were triggered by a human's decision is to fire all your humans, please don't do that. (Are you a robot overlord?) Humans, for all our flaws, are still the most capable tool for solving technical challenges. Engineers who set the figurative fires that cause incidents are also the best firefighters for solving the problems.

The most thorough answer the academic world has formed to respond to engineering mistakes is human factors, also referred to as *ergonomics*, which is the study of human psychology and physiology in design. This field of study applies knowledge about humans from many disciplines — psychology, sociology, user experience, engineering, industrial design — and enables people to design better products and systems, all with the main goal of reducing human error.

TECHNICAL STUFF

If you're thinking, "I thought ergonomics had something to do with my chair," you're right! Physical ergonomics is what improves the products you use every day, from your chairs to your computer screens. What you should be concerned about in DevOps is cognitive ergonomics and organizational ergonomics:

NOOPS

Occasionally the term *NoOps* gets thrown around in the DevOps space. NoOps doesn't mean a lack of operations engineers. Instead, it indicates a focus on automating everything related to operations — deployment processes, monitoring, and application management. Whereas DevOps focuses on helping developers and operations folks work together more seamlessly, NoOps aims to prevent developers from ever interacting with an operations engineer.

In many ways, NoOps is a part of the DevOps movement, but with a different approach. DevOps focuses on people, processes, *and* technology. NoOps relies on specific software solutions to manage things like infrastructure and deployment pipelines — solving only the technical challenges we face. You can think of DevOps as a fully encompassing cultural shift and NoOps as a much more narrow technical solution.

I am not personally a proponent of NoOps because the skills and experience of operations professionals go beyond manual deployments and other processes primed for automation. Operations engineers are the most qualified for automating toil (or rote work) but also for architecting systems with complex infrastructure components in mind.

- **Cognitive ergonomics** is the study of how humans perceive and reason about their environment. How do people make decisions or react to certain stimulus? What makes one person extremely reliable and another flaky?
- **Organizational ergonomics** is the study of systems and structures inside organizations. How do teams communicate and work together? What makes some teams cooperative and others competitive?

Focusing on systems: Automating realistically

Unfortunately, deploying changes to the human brain is still something you may struggle to accomplish. Instead of focusing on preventing humans from making mistakes — an impossible task — DevOps processes recommend that you turn your attention to creating and implementing automated systems along the entire development process.

Automation is the best-known way to combat human error. If you asked humans to write their name a million times in a row, they would eventually misspell their names. Their own names! (They'd also develop a repetitive stress injury.) But if you asked a robot to complete the same task, it would accomplish the job flawlessly, identically printing a name a million times, without error.

The same concept applies to your applications. If asked to repeat rote tasks, humans will make mistakes. Four areas primed for automation:

- **Code:** Software developers design and build solutions via code. Developers manage their source code and often work on the same portion of a codebase simultaneously.
- **Integration:** Code changes must be merged from multiple developers into the master branch of a code repository.
- **Deployment:** After being merged, the code must be deployed. This can often mean releasing updates, changing configurations, and even deprecating services.
- **Infrastructure:** An application must be run on hardware. Depending on the updates to code, infrastructure may need to be instantiated, provisioned, or terminated.

The automation tools in each of these spaces experience a quick rate of churn. Don't be surprised if your beloved solution loses favor a year or two from now. Tech will always have a "hot new technology" that everyone's talking about, but don't be distracted by the latest new thing. Focus instead on the best solution for you and your team regardless of how popular the tool is.

Embracing the best solution for your team is always the best answer. That said, sometimes you do find some benefits of moving with the crowd:

- **Popular tools often have the best documentation and answers on technical forums.** The more people that use a project, the more likely someone is to have documented the code, built a demo, created an instructional video, or answered questions on forums like StackOverflow, a website with nearly endless answers to technical questions. Popular tools are also likely to be accompanied by published documentation and examples. I encourage you to read the docs of any tool before you select it because the tool you choose will determine how smoothly your development goes as you move forward.
- **Popular tools are often open source software (OSS).** Open source software is a broad term to describe tools that are (usually) free to use and open to community input. You can actually go into the tool's source code, implement a

change, and submit a request for the change to be approved. OSS communities are often run by a small team of volunteer engineers. OSS has many benefits, but in this case, you can actually tailor the tool to you. You can clone the current code and build a tool on top of it, or you can commit your code to the project and help others solve the same problem you're solving. Read more about integrating with OSS in Chapter 19.

Using automation tools to avoid code integration problems

The more automated monitoring and responses you can build into your incident management, the less you'll have to depend on human escalation and resolution.

Before you can automate any type of incident response, you must identify the key metrics that you want to monitor. Obvious choices might include availability, initial response times, uptime, traffic, and revenue. Others might also add SSL expiration, DNS resolutions, and load balancer health checks. Many of the granular metrics that your team monitors and responds to will be based on your company's key performance indicators (KPIs).

The best things to automate are processes your engineers manually engage with regularly. Configure your monitoring tools to inject relevant information into your alerts. Status pages are fantastic tools for updating stakeholders at regular intervals. You can build slash commands into chat tools to automatically update your status page. Finally, don't forget about automating data collection. Logging tools can help you identify what went wrong on a diagnostic level as well as what was impacted. In hindsight, you'll be able to better understand which areas of your application and infrastructure are brittle and what action you need to take to prevent similar incidents in the future.

Following are some automation tools that you and your teams can use to mitigate incidents at every stage of development. These tools are handy, but you should never rely solely on them to solve the challenges your team faces. Tooling will never remove the need to build a culture, processes, and systems that avoid human error.

» **CircleCI:** A cloud alternative, CircleCI supports many mainstream languages and offers up to 16x parallelization. It is container based, so pricing is based on the number of containers you use. Circle is one of the fastest (and most expensive) options.

- **Jenkins:** Written in Java, Jenkins is open source and extremely flexible. The Jenkins plug-in list is lengthy, to say the least. The learning curve can be a bit steep but is definitely worth the time. You can control Jenkins via the console as well as a graphic user interface (GUI).
- **Go CD:** Like Jenkins, Go has mastered pipelines to help you implement continuous delivery. Its parallelized execution eliminates build bottlenecks. Go is completely free and offers paid support.

You likely already use some kind of source code management or version control tool like Git. In fact, these tools are so ubiquitous that you probably don't think of them as automation. But they do! Imagine if your engineers had to merge code manually. It'd be a nightmare.

Even if your team hasn't yet adopted Git (don't stress!) you may use something like SVN or Mercurial. Whatever the tool, it enables you to manage the work of multiple developers who are making changes to the same codebase. Such tools, make it relatively easy to visualize the differences between two branches, choose the most recent changes, and merge them into one branch — usually the main branch called trunk or master. (I said relatively; don't curse my name the next time you have a merge conflict.)

I highly recommend adding a continuous integration (CI) tool to your toolset as well. You can consider some of the tools that follow as deployment tools. In fact, most of the tools mentioned in this book are difficult to classify into only one category because they span a number of areas. For this book, I highlight and categorize tools based on their core competency — the feature for which they are best known.

Handling deployments and infrastructure

When it comes to application deployment and configuration management, the available tools aren't always familiar to people and often require some degree of integration into your current infrastructure and deployment processes. Examples include Ansible, Chef, Puppet and Salt, although this list is far from exhaustive.

As infrastructure becomes exponentially more complicated, observing your systems in real time becomes ever more difficult and the importance of automation in deployments (and infrastructure) increases.

- **Ansible:** Written in Java, Ansible is a Red Hat suite of DevOps-focused products that help teams deploy applications and manage complex systems. Ansible attempts to unify the teams of developers, operations, quality assurance (QA), and security as well as to simplify their repetitive tasks.

- **Chef:** Bridging the gap between engineers and operations folks, Chef is a leader in the continuous automation space. Chef can manage up to 50,000 servers by turning infrastructure configurations into code.
- **Puppet:** Puppet products seek to deliver real-time information about your infrastructure, automate tasks driven by models *and* events, and create continuous integration and continuous deployment (CI/CD) pipelines that are easy to set up. Puppet helps teams support traditional infrastructure as well as containers.

Limiting overengineering

Imagine two bakers. One produces a perfectly warm and airy loaf encrusted by a crisp exterior. Breaking it releases the irresistible, yeasty smell of fresh bread carried by just a touch of steam. The other produces a dense, dry bread encased by a rock-hard crust. Yuck. The bakers followed the same recipe and used the same ingredients. So what went wrong?

In the latter case, the baker overkneaded the dough. The overworked gluten produced a dense, unappealing product. Although both loaves might be equally nutritious, eating the second loaf would be more like gnawing on a rock than biting into bread.

ALL COMPANIES ARE TECH COMPANIES

Whether you want to admit it or not, your business is in tech, which can be hard to internalize. Here's a story that illustrates the point. Recently I went to LabCorp, which is a company that draws blood, evaluates the sample against various tests, and sends the results to your doctor. Not very technical, right? Only when I went in for my blood work, LapCorp's coding system was down nationwide. LabCorp had no analog redundancy. Unless the technicians and phlebotomists had memorized the specific code that they're required to put on a blood sample for processing, they could not see a patient.

This situation meant that almost all customers had to leave and return another day, which was a lot of lost business. Yet, you might have been tempted to not characterize LapCorp as a tech company. Tech enables all companies to scale their services to more customers than would be possible without it, but the very tech you depend on every day will occasionally fail you. The truth is that you don't have the option to pretend you're not a tech company, no matter what business you're in.

Code isn't all that different from making bread. Styles vary but most recipes require the same basic ingredients and follow one of a handful of formulas. More often than not, the simplest solution is the best. But no matter how many great ideas you come up with, you'll have some fairly terrible ones as well. The trick is to recognize the terrible ones quickly and invest heavily in the great ideas. Discerning the difference is a learned skill.

An engineer loves few activities more than, well, engineering. Engineers love solving problems. The more complex, the better. Upon hearing about a problem, most engineers want to immediately dive into the first solution that pops into their head.

This instinct, although admirable, doesn't always lend itself to finding the best solution — only the most obvious one. Often when you hear the term *overengineering,* the reference is to code that's overworked or solutions that are unnecessarily verbose or complex.

Here are a few warning signs that a solution is overengineered:

- **The problem is more easily managed manually.** Not every problem needs to be automated. Do you need to write a to-do app when pen and paper work just fine? Maybe, but probably not. Make sure that a technical solution is efficient and necessary before developing it.
- **The code is unusually verbose.** If the lines of code required to solve something are double the amount needed for typical bug fixes and feature implementations, look into why.
- **The solution wasn't peer-reviewed.** All implementations should be discussed with a peer prior to development or reviewed by a peer before being merged into the rest of your source code. This prevents myopic and unnecessary code.
- **The code is difficult to understand.** If a junior engineer can't interpret what a piece of code is doing within an hour, take that as a warning sign. Code must be maintained, and all engineers need to ensure not only that their code works but also is readable by their colleagues and their future self.
- **A free or cheap tool exists that solves the problem.** Spending time engineering a solution to a problem that has already been solved is foolish. Research the tools that already exist to ensure that writing code is necessary.

Before you automate something, solve the problem manually first. Even if it requires — gasp! — pen and paper or, arguably worse, a spreadsheet. Making sure that your approach works before you automate it is important. Otherwise, you end up wasting time and engineering resources on unused, ineffective solutions.

Humanizing On-Call Rotation

Being on call is akin to being available to handle emergencies. If the site goes down or your customers are impacted by a technical failure, you are the designated person to manage the issue — no matter when it happens.

Imagine that you have to rush your toddler to the ER at midnight because he decided to swallow your wedding ring. The on-call surgeon affiliated with the hospital might be paged to come in and treat your child. They are physically close to the hospital and prepared to go in when necessary. You can apply the same principle to on-call engineers in a DevOps organization.

When on-call duties become inhumane

One of the most significant cultural and organizational shifts in adopting DevOps revolves around a shared on-call responsibility. Traditionally, developers would write the code to implement a feature and pass it to the operations team to deploy and maintain. This meant that only a handful of operations engineers were on call for when a poorly developed piece of code failed.

Having too few people on call is one of the key problems DevOps attempts to solve. By sharing responsibility, both teams can have autonomy and mastery over their work. That shared responsibility also means that the burden of being on call is distributed over a much larger group of people, which prevents burnout.

Site reliability has become increasingly important. Many companies lose hundreds of thousands of dollars for every hour their sites are offline. Companies can build resilient systems to avoid catastrophic failure, but every company must also keep engineers on call to handle unexpected emergencies.

The typical process for responding to an incident looks something like this:

1. Customers are impacted. Maybe your monitoring software has alerted you that the site's taking 20 seconds to load. Maybe there's a regional outage and European customers are yelling at you on Twitter. The types of incidents are nearly limitless, but someone's mad.
2. The primary person on call is alerted. Services like PagerDuty and VictorOps allow you to customize who gets alerted and how. If the primary person on call does not respond within a set amount of time, the secondary contact is paged.
3. An engineer attempts to fix the problem. Sometimes the issue isn't critical enough to address in the middle of the night and can be fixed the next morning. Other times, the server room is literally flooding and someone needs to get a bucket. (Hurricane Sandy in 2012 flooded two major data centers in lower Manhattan.)

This all sounds great. Sites stay online and responsibility is shared, right? Not usually. Unfortunately, being on call can quickly become inhumane. Traditionally, system administrators and operations engineers are the only folks who end up on call, which goes against core DevOps principles and reinforces silos. I believe strongly in shared responsibility. You build it; you support it.

Humane on-call expectations

Making a true jump to a DevOps model for creating, deploying, and supporting sites requires that on-call duties be shared by every engineer involved in a product. On-call rotation is an opportunity, not a punishment. It's an opportunity for engineers to think differently, learn new skills, and support their team and for the organization to build better systems and processes to:

- Document code better
- Create runbooks (step-by-step guides of what to do) for common issues that still require manual work
- Empower individuals to ask questions and take risks

Developers who are empowered to support their own code build better products, period. These developers begin to think about their code in terms of reliability and resiliency *while* they develop, rather than as an afterthought, if they think about those aspects at all.

When you're on call, you're expected to be available to respond to any incidents that may arise. Some folks split workdays into on-call shifts. For example, Tim is on call from 8:00 a.m. to 10:00 a.m. every morning. Others cover nights and weekends on a rotating schedule. If this approach works for you and your team, go for it!

Based on my experience, I suggest something a little different. People do their best work when they have extended periods of time away from being "on," and that means having full days without having to worry about being paged.

In 2010, LexisNexis conducted a survey of 1,700 office workers in several countries. The study found that employees spend more than half their day receiving information rather than putting that data into practice. Half the respondents said that they were approaching a mental breaking point from being overwhelmed with information. Breaks are a critical aspect of productivity and work-life balance.

Figures 17-1 and 17-2 show some example schedules. Figure 17-1 shows how two people can share daily, on-call duties while keeping at least three clear days in their week. Figure 17-2 divides the duties among four people. Each is required to be on call at least one day per week but no more than three days per week. Each shade represents a different person. The columns are days of the week and the rows are weeks (four rows represent a typical month).

REMEMBER

Each person is on call from 5:00 p.m. to 5:00 p.m., which is simple if you're all in the same office. If your organization is remote-first or remote-friendly, you need to choose a single time zone for everyone to follow to ensure 24/7 coverage.

On-call rotations come many forms. The examples I provide are intended to help you get going, not limit you. You should tailor the schedule to make it work best for your team. If you have a globally distributed team, you can adopt a follow-the-sun rotation that puts engineers on call during normal business hours before they pass the responsibility to those working normal business hours in a different time zone. Find the days, times, and frequency of on-call rotations that can balance incident management with humane on-call practices.

	Sunday	Monday	Tuesday	Wednesday	Thursday	Friday	Saturday
W1	Tim, Ops	Ann, Dev	Ann, Dev	Tim, Ops	Tim, Ops	Ann, Dev	Ann, Dev
W2	Ann, Dev	Tim, Ops	Tim, Ops	Ann, Dev	Ann, Dev	Tim, Ops	Tim, Ops
W3	Tim, Ops	Ann, Dev	Ann, Dev	Tim, Ops	Tim, Ops	Ann, Dev	Ann, Dev

FIGURE 17-1: An example of a two-person on-call schedule.

	Sunday	Monday	Tuesday	Wednesday	Thursday	Friday	Saturday
W1	Tim, Ops	Ann, Dev	Mel, Ops	Don, Dev	Tim, Ops	Ann, Dev	Ann, Dev
W2	Ann, Dev	Mel, Ops	Don, Dev	Tim, Ops	Ann, Dev	Mel, Ops	Mel, Ops
W3	Mel, Ops	Don, Dev	Tim, Ops	Ann, Dev	Mel, Ops	Don, Dev	Don, Dev
W4	Don, Dev	Tim, Ops	Ann, Dev	Mel, Ops	Don, Dev	Tim, Ops	Tim, Ops
W5	Tim, Ops	Ann, Dev	Mel, Ops	Don, Dev	Tim, Ops	Ann, Dev	Ann, Dev

FIGURE 17-2: An example of a four-person on-call schedule.

Managing Incidents

In my talk "This Is Not Fine: Putting Out (Code) Fires," (`https://www.youtube.com/watch?v=qL2GFB3mSs8&t=69s`), I speak a lot about incident management and how it relates to another type of firefighting — the kind with actual flames. Engineers and operations pros can take a lesson from the way firefighters prioritize how they combat incidents that are way more dangerous than tech failures and apply those steps to addressing incidents. (See the sidebar "Putting out code fires" for more info on how firefighting principles can work in tech.)

PUTTING OUT CODE FIRES

After a series of extremely destructive and deadly wildfires in California during the 1970s, a task force called the Firefighting Resources of California Organized for Potential Emergencies (FIRESCOPE) was formed. FIRESCOPE distilled its findings into four priorities that you can also use when approaching incident management:

- Flexibility
- Consistency
- Standardization
- Procedures

These principles have helped fire departments around the globe to consistently address the broad number of incidents they're required to handle — from rescuing ducklings from a drain to rescuing people from a burning high-rise — using proven procedures. As you make the move into a DevOps culture, your team's success will hinge on the same principles when addressing incidents.

The incidents you deal with in tech can sometimes be as simple as an odd user interface bug in a drop-down list, which isn't exactly life-threatening or worthy of a hotfix at 4 o'clock in the morning. Sometimes, though, your software goes wrong in spectacularly terrible ways. For example, in 2003, a performance issue in utility software caused a blackout in the American Northeast. And in 2000, radiation therapy software in Panama failed to account for a workaround used by doctors, resulting in eight patient deaths and another 20 radiation overdoses.

These situations are vastly different from simple bugs and performance issues. Yes, a slow site loses money and causes customer disruption. But having people express anger at you on Twitter is much less stressful than having people die or watching your company go bankrupt by the minute.

Making consistency a goal

If you've ever flown in a private plane, you know how much pilots love checklists. Well, maybe they don't *love* them, but they certainly use them. Checklists are a big part of why air travel is by far the safest way to get from point A to point B.

For pilots, these checklists are part of a preflight flow that checks switches, circuit breakers, and emergency equipment. Pilots run through this process before every flight, with no exceptions. This consistency moves the process beyond regular consciousness and into muscle memory. Pilots with even just a few years of experience don't need to think about their preflight flow; it's automatic.

Along the same lines, you should create an incident checklist for your team that your team will automatically follow when it's needed. If you're not sure what to include, start with these actions:

- **Notify appropriate colleagues.** Depending on who is involved in the incident, keep up-to-date contact information for everyone on your team.
- **Deploy a status page.** Inform customers what service or features are affected. Be sure to include the contact information for your support team and the time of the last update.
- **Rate the incident.** Your checklist should include clearly defined severity ratings to help the first responders appropriately escalate an incident to legal or executive management.
- **Schedule a post-incident review.** Post-incident reviews are a key part of reducing human error and building resilient systems. How else do people learn if not through mistakes? If possible, schedule it within 36 hours of the incident.

Adopting standardized processes

The more you standardize your emergency preparation, the more people you can rely on to step in and help fix the problem. If only one person can address a certain issue, that person becomes a single point of failure, which is absolutely unacceptable in modern tech companies.

Make the checklists and incident response protocols available to everyone on your team — even the folks who aren't on call. Making them available to everyone ensures that the entire company is on the same page and eliminates needless questions from teams like customer support during an incident.

To fully adopt DevOps practices, developers must store the source code in a place that the ops teams can access. Also, give developers access (at least read-only) to all logs and machines. This approach enables both sides to dig into all areas of the tech — source code and infrastructure — without asking for permission. The alternative is to rely on people from other teams to be couriers of information — a time-intensive and inefficient process.

Establishing a realistic budget

The roots of many of the popular trends in tech are in large companies that adopted a certain tool or practice. For example, site reliability engineering wasn't a well-known concept or role until Google published *Site Reliability Engineering: How Google Runs Production Systems*. React, a JavaScript library, took off in popularity largely because Facebook developed and promoted it.

Your company may not have the financial resources of companies like Microsoft, Google, Amazon, and others, so your incident response procedures need to be designed with a budget in mind. Monitoring every service is impossible. Instead, focus on the ones that your company uses the most frequently or that have the greatest impact to your customers. I strongly recommend centralized logging to create a way for logs to be captured at increasingly larger intervals as time goes on. In other words, find a balance between visibility and budget in storing log data and performance metrics.

LESSON LEARNED: NETFLIX RESILIENCY

In February 2017, Amazon's S3 (web-based storage) experienced widespread issues in the US-EAST-1 region. It effectively brought down much of the Internet — including Amazon's own status page. Netflix was one of the only major websites not to experience any issues, even as a customer of AWS.

Netflix, it turns out, had learned this lesson five years earlier when a storm knocked the site offline for about three hours. In the post-incident review, Netflix realized that it was vulnerable to regional outages. A company like Netflix loses hundreds of thousands of dollars for every hour of downtime, if not more.

The solution, for Netflix, is to switch availability zones in AWS automatically when one goes down. Users will never be affected by regional service disruptions. This solution is expensive, however, because you can never max out capacity and balance your users efficiently across zones. Otherwise, you wouldn't have the volume available to move users to another region when one fails.

The costs for this type of solution add up quickly and are prohibitive for many companies. Budget constraints are a vital piece of your overall strategy and should inform many of your decisions.

Making it easy to respond to incidents

Incident management protocols must be generic enough to respond to events with varying levels of urgency and importance. They should also maintain clear procedures for people to follow while they're rubbing sleep from their eyes in the middle of the night and trying to wrap their brains around the problem. Following are a few tips that can help your engineers master incident management:

- **Make it easy and acceptable to escalate.** You're better off overreacting rather than underresponding to a situation. The primary person on call should be able to page the secondary engineer on call without retribution.
- **Use a single communication tool.** When different teams within an engineering organization use multiple communication tools, absolute chaos during an incident can ensue. Engineers must be on the same page, and being able to scroll back through conversations or reach a colleague quickly via a video conferencing tool is essential. Always use the same medium to reach your coworkers. I highly recommend using a chat app like Slack or hopping on a set incident video conferencing call via a tool like Zoom.

  Every method of communication comes with pros and cons. Using video calls to communicate during an incident creates a more fluid experience for the engineers on call but limits your ability to include that information in the post-incident review. Group chats, such as Slack, aid you in better capturing the timeline of an incident response but may create confusion for the engineers responding. (Messages written in haste tend to be short and lack the detail and context that you could provide verbally in a fraction of the time.) Two compromises exist: Record the video calls or have someone summarize events for the group in a written format.
- **Standardize the initial investigation.** Create a step-by-step list so that any engineer can quickly begin to triage a situation. Is there a widespread AWS outage that's causing half of the Internet to go down? If not, monitoring tools and logs will be your best bet to home in on the problem. Only if all else fails is it appropriate to allow engineers to "sniff test" the issue and follow their gut.

Cloud computing services like AWS and Azure host multiple locations around the world. Every location is composed of regions and availability zones. A region is a geographic area. AWS has US-EAST-1 in Northern Virginia and AP-SOUTHEAST-1 in Singapore are considered regions, for example. Multiple availability zones exist within each region.

Urgency is not the same as *importance*. The distinction between these two qualities comes into play when you are discussing on-call procedures. Urgency defines how rapidly something must be resolved. The site's down? That's pretty urgent. Customers can't make purchases? Also urgent. A rarely used API is failing gracefully? Not urgent. Important, but not urgent.

Important incidents that lack urgency can wait until the morning when an engineer can give their best effort to fix the issue. Making this simple distinction will save your team from buggy fixes and prevent your engineers from becoming needlessly burnt out.

Responding to an unplanned disruption

In any situation, it's always best to assume the worst. As mentioned in the previous section, escalating a situation and treating it as a more severe incident is always better than underreacting.

Also, decisions should be made quickly during a crisis. Hierarchy is always going to be a controversial topic in tech. But, especially when responding to incidents, I recommend a strong response hierarchy with designated roles. Your team should include an incident commander (IC), a tech chief, and a communications chief.

Different resources include various version of the number and type of incident roles. You may hear things like first responders, secondary responders, subject matter experts, and communication liaisons. I choose to focus on the three I've listed because they cover the three most important roles of an incident response: someone to make decisions, someone to lead engineers in the technical response, and someone to record the details of the incident. Feel free to experiment with your incident-response procedures and find what works best for you and your organization.

Think of the primary person on call as the first one on the scene. They will not necessarily be the person most equipped to handle the particular issue. In fact, the person doesn't have to be an engineer at all. The primary person on call is simply the person who triages the issue. This person is tasked with assigning a degree of *urgency* to the alert.

Make sure that you rotate incident teams, just as you do in your on-call rotation. Rotating teams enables people with different skills and interests on your team to become proficient — and more confident — in other areas. Every person on your team should have the opportunity to be trained and serve in each role. Figure 17-3 illustrates an incident response hierarchy. The incident commander will oversee and provide resources for the tech chief and the comm chief, including supplying them with the appropriate number of engineers to assist (represented by the small boxes beneath each chief).

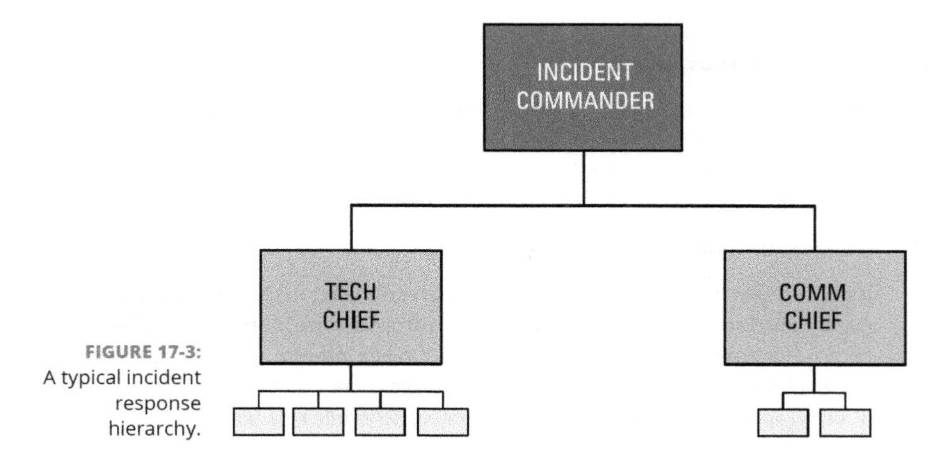

FIGURE 17-3: A typical incident response hierarchy.

You can see how this hierarchy is put into action in the following steps, which outline the procedure for handling an unplanned disruption.

1. **Make an initial assessment.**

   At the start of an incident response, the IC begins sizing up the situation. Be sure to categorize and prioritize the incident. Categorization doesn't have to follow a particular pattern, but your classes of incidents should enable you to group similar incidents and evaluate trends. Prioritization is centered around urgency. Is this customer-impacting? How wide-spread is the incident? How many engineers might be required to help fix it? The IC determines how many engineers the tech chief needs to notify.

2. **Communicate during triage.**

   I suggest hopping on a video call to discuss the disruption. Zoom and other video conference tools help you communicate in real time. Although Slack and other messaging tools have become part of everyday communication, the power of face-to-face communication, especially during a crisis, is critical. Your engineers need to communicate with each other verbally while their fingers are busy logging into machines or digging into code. If you opt for a messaging tool like Slack, you'll be able to include that transcript in the post-incident review. If you triage on a video call, be sure to designate one person to record who said what and which solutions were attempted.

TIP

   A societal norm exists for the women you work with to default into administrative or non-technical roles. You can see this in who most frequently ends up being the person to record the conversation or serve as comm chief. Be sure to watch for this gender-biased default and counter it by ensuring that engineers who don't identify as male also serve as incident commanders and tech chiefs.

3. **Add engineers as necessary.**

   After you dig into the incident, you may realize that you need a subject matter expert who is particularly equipped to deal with the type of incident you're experiencing. They could be deeply trained in the particular tool or technology, or they may be the engineer who implemented a specific function.

4. **Resolve the issue.**

   It's easier said than done, but the engineers responding to the incident will eventually discover the steps necessary to restore service. At that point, the comm chief can relay important information to key internal and external stakeholders, the IC can schedule a post-incident review (if they haven't already) and the tech chief can help engineers schedule rest and recovery before the post-incident review.

LEARNING FROM THE MISTAKES OF OTHERS

You're never the first to fail. Even when it feels as if you're the only one who could have fallen on your face in such a spectacular and unique way, you're not, I promise. Although tech isn't new, it has reached saturation in the developed world, which means that you have plenty of resources improve your avoidance of and approach to incidents.

In January 2017, GitLab, a git-repository hosting service and manager, experienced a site outage because of the accidental removal of primary database. GitLab was down for *18 hours*. That's enough to give any engineer heart palpitations. To its credit, the company was extremely transparent about the event, going so far as to keep notes in a public Google document and livestream its recovery on YouTube. The full post-incident review of the event as well as the data loss outcomes are well worth the read.

Ultimately, GitLab discovered that it had two problems:

- **GitLab.com had an unplanned disruption after the wrong directory was removed.** The primary database directory rather than the intended secondary database directory was removed. Replication stopped because of a spike in load. Restoring database replication, after it was stopped, required a manual process that was poorly documented, in this case.
- **Restoring the site required a copy of the staging database.** This database was stored on a slower Azure VM. Disk snapshots weren't enabled and attempts to back up the database failed silently because of a PostgreSQL versioning issue.

Could there be a more perfect storm? Here's my favorite part of the published postmortem: "Why was the backup procedure not tested on a regular basis? Because there was no ownership, [and] as a result nobody was responsible for testing this procedure."

This cascading series of failures could affect any organization. No one is exempt from those unknown unknowns of unplanned downtime. The distinctions between organizations that let failures overwhelm them and those that use the same incidents as learning opportunities are attitude and preparedness.

GitLab was brutally honest with its customers and has since improved its recovery procedures. Many companies would have revoked production privileges from engineers, thus creating a bottleneck. Instead, GitLab made it more obvious for engineers which host they're using.

The worst incident response I've ever witnessed was so traumatic that I don't even remember what went wrong. I distinctly remember how it transpired, however. A major disruption of service brought a small startup down for hours, and sometime in the early morning, the CEO called the two most senior engineers — who, at that point, had been troubleshooting without a break for half a day. I've never heard a man make so many threats over the phone. The CEO assured the engineers that the situation was their fault and promised that if they didn't fix it soon, not only would he fire them, he would make sure that they were never hired at a venture capital-funded startup ever again.

Put yourself in the shoes of those engineers. You've been working for hours. You haven't eaten. You've barely had time to make coffee, which is the only thing keeping you moving at this point. The CEO's blame and threats were enough to put anyone in a state of panic, which is about the worst thing you could do to engineers who are working on fixing the issue.

That CEO made a critical error: Distracting the engineers who were working so diligently for him. He took their attention away from the emergency they were triaging and put their attention on their future. No matter how chaotic and stressful an incident becomes, always remember that the folks working to remedy the situation are doing their best and care about fixing the issue as successfully and quickly as possible.

Every organization will experience a major incident at some point; it's inevitable. But how you prepare for those incidents and cope with them in the moment is what separates teams who embrace DevOps and those who don't.

Empirically Measuring Progress

More and more companies are beginning to develop a DevOps culture and implement change within their organizations, yet most don't measure incident response. In fact, most companies don't even know which metrics matter. Success in incident management doesn't go from zero to perfect, and achieving it is hard. But the best way to improve your success is to start gathering and analyzing metrics. This section provides some metrics for you to start observing and tracking. If you're just getting started, now is not the time to start setting goals or adding these measurements to personnel reviews. Instead, think of them as single points of data that together paint a broader picture of your company's success.

REMEMBER

I want to be clear about one thing. I've chosen to put this information as the last part of this chapter for a particular reason: It's the least important. The metrics in this section are simply data points that serve as the foundation of a larger organizational conversation. These are never meant to be the only measure of success. Instead, track them as a way of measuring the progress of your team as they continuously improve their incident management.

Mean time to repair (MTTR)

The mean time to repair refers to the average time your business is impacted during incidents. When collecting this metric, also include latency, the time from when the failure first occurred to when it was detected. You likely calculate latency after the incident is resolved so that you can reasonably estimate, via logs and other data, when the failure began to impact the affected service before an engineer realized it was a problem. The formula looks like this:

MTTR = total time of impact / number of incidents

People also sometimes use MTTR to describe the mean time to recovery, the amount of time your team takes to resolve an issue as well as mean time to respond, or the time an organization takes to acknowledge and initiate a response to a problem. (Remember, a mean assumes normal distribution, and an 18-hour outage like GitLab experienced will exaggerate their response time. MTTR is just one data point.)

Mean time between failures (MTBF)

In short, MTBF is the average uptime for a service between incidents. The higher an organization's mean time between failures, the longer the service can be expected to work without interruption. Here's the formula:

MTBF = total uptime / number of incidents

Although MTBF can provide a helpful piece of data, many DevOps organizations are moving away from tracking MTBF because failures simply can't be avoided. You could instead track customer-impacting incidents (rather than service failures of which the user is never aware).

Cost per incident (CPI)

The cost per incident is simply how much money your company lost because of the service interruption. This calculation has two phases. The first is how much the actual incident cost you: Were customers unable to make purchases? The second is the cost of bringing your services back online: How many engineers were required to address the issue? Here are the formulas:

> Lost revenue (LR) = average revenue * time
>
> Cost to restore (CR) = number of engineers * average hourly salary * time
>
> CPI = LR + CR

CPI adds up fast. You can use these calculations to convince even the most stubborn executives to put resources toward preparing for incidents, paying down tech debt, testing more rigorously, and improving application security.

DevOps Research and Assessment (DORA) goes further than CPI and calculates the cost of downtime using the following formula:

> Cost of downtime = deployment frequency * change failure rate * mean time to recover (MTTR) * hourly cost of outage

You can read more about calculating your cost of downtime at `https://victorops.com/blog/how-much-does-downtime-cost`.

» Stepping through the phases of an incident

» Reviewing contributing factors in post-incident reviews

Chapter 18
Conducting Post-Incident Reviews

Engineers are much more practiced at reacting to incidents than they are to proactively preparing to manage and avoid them. Post-incident reviews aim to empower engineers to look at the causes of an incident, the steps taken while responding to an incident, and the steps necessary to avoid a comparable incident in the future.

People used to refer to post-incident reviews as postmortems, and you can still find a lot of valuable information if you search for this term. However, the word is a bit morbid with its connotation of death. For most software engineers, outages mean inconvenience to customers and loss of company money. Few engineers deal with life-and-death situations in the use of their products, and keeping that perspective in mind when addressing failures is important.

In this chapter, you dive into the contributing factors of failure (going beyond root cause analysis), the phases of an incident or outage, and the way to run a post-incident review.

Going beyond Root Cause Analysis

If you've been in tech long enough, you've heard the term *root cause*. Looking for the root cause meant to identify the single source of failure in an incident. The problem with root cause analysis — and why it's not typically used in modern operations teams — is that a root cause almost never exists. It's the same as the trope, "It's always the last place you look!" Well, yeah, you found the thing. You're not going to keep looking. In his 2017 PuppetConf talk, "The Five Dirty Words of CI," J. Paul Reed noted, "What you call a 'root cause' is simply the place where you stop looking any further."

Unlike simple, linear systems, the code and infrastructure you operate and maintain are incredibly complex. A single "root cause" simply does not exist. But rewind to the days of waterfall processes and uncomplicated monolithic architecture. In those systems, root cause analysis made more sense. You could view the system as a whole and pick out the piece along the process that failed. Changes were more infrequent, and root cause analysis was a way of thinking through risk.

The systems you operate are no longer simple, likely aren't monolithic, and are typically a mess of legacy code, new additions, multiple languages, unknown dependencies, and a cordoned-off section of obfuscated code written in ColdFusion that works — though no one knows why — and has been converted into an "engine" that powers the central portion of your user-facing features. Sound about right? Of course it does. I've never met a codebase older than two weeks that's neat and tidy. Humans are messy, and humans write code; therefore, code is messy. You function within constraints and your code expresses symptoms of those constraints, whether those symptoms are related to finances, safety, or time.

The industry has moved beyond root cause analysis, and the time has come for you to replace it with a more worthwhile process of review. You can actually have a complex monolith, and a fair amount of companies continue to use monoliths successfully. Modern architecture does not require that you rewrite your entire system to be completely compartmentalized. It does, however, require you to think about the pros and cons of each decision and recognize that every move is a decision, even if the decision is to *not* take action.

Figure 18-1 compares the general complexity of monolithic and microservice systems. You can see the increased complexity of the microservice system in the hundreds, if not thousands, of connections in a microservice architecture.

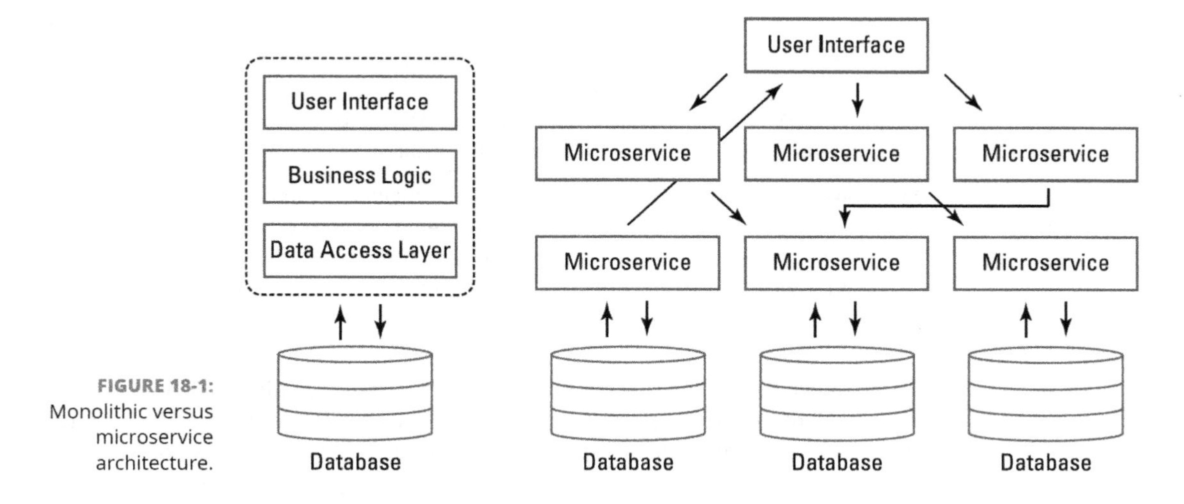

FIGURE 18-1: Monolithic versus microservice architecture.

Stepping through an Incident

Broadly speaking, incidents can be broken into five steps: Discovery, Response, Restoration, Reflection, and Preparation, as shown in Figure 18-2. The purpose of breaking an incident into different phases is to better understand each step of the unplanned work:

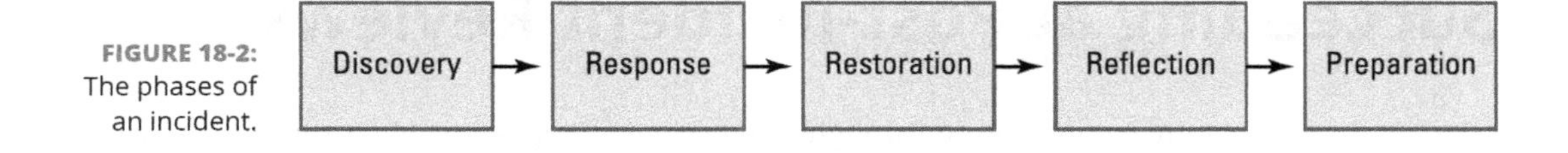

FIGURE 18-2: The phases of an incident.

- **Discovery:** This phase starts when the issue is detected. Services can be impacted for a period of time before you realize it.
- **Response:** This phase is the scramble of trying to determine the source of the issue. Was it a recent deploy? Is the service that's down the source of the issue or could the cause be an ancillary service? Does the have a code problem or is your infrastructure failing?
- **Restoration:** At this point, you've identified the issue and are working on solving it. This phase is often one of the shortest ones of the incident. After you know what's happening, you usually discover a straightforward fix, even if it means rolling back a deploy or reverting to the last issue-free build.

» **Reflection:** This phase is where a post-incident review takes place. You and your team get together within 48 hours of the incident and discuss the process. What went well? What went poorly? What work needs to be done to prevent the same type of incident in the future?

» **Preparation:** During the preparation phase, engineers complete the work determined necessary during the post-incident review. You should assign the work to an engineer who can see the process completed as well as set a due date for when the work should be done. Just be sure to clear enough time in the schedule so that the work *can* be done.

Whereas most teams put the most work into the first three phases of an incident, the last two often get forgotten because the urgency falls away as soon as service is restored. Figure 18-3 depicts the phases of an incident that focus on the post-incident review: Reflection and Preparation. The review should occur during the reflection phase, but the work that review determines to be necessary will be completed during the preparation phase.

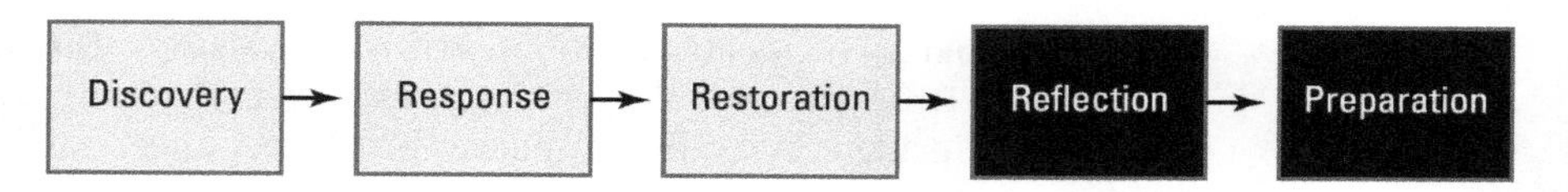

FIGURE 18-3: The phases that focus on post-incident reviews.

Succeeding at Post-Incident Reviews

Many companies go through the motions of a post-incident review without fully taking advantage of the entire process. (Even more companies don't bother with it at all, which is a massive mistake.) If you use the fundamentals of a post-incident review listed in this section, your team will be equipped to evaluate past mistakes and prepare for future failure.

Scheduling it immediately

Schedule the post-incident review *while* the incident is happening. You may feel as though you're scheduling a dentist appointment while your house is burning down, but the purpose is to acknowledge that something has gone wrong and a discussion will take place to prevent it again in the future.

You should hold the post-incident review no more than three days after you've resolved an incident. Ideally, you hold the review within 36 hours. The human brain is fickle and notoriously terrible at retaining detailed information. The sooner you get together to discuss the incident, the more valuable the meeting will be.

Including everyone

Put the scheduled post-incident review on a shared company calendar so that everyone can see it. It's critical to include the first responders and those who were directly involved in the incident in the post-incident review. But don't stop there; open it up to everyone. I can't think of a better way to help other departments understand the challenges of engineering than to invite them to a post-incident review. If you have embraced the practices of healthy post-incident reviews, opening the review to everyone creates a wonderful opportunity to educate others. Just be sure that your team is ready to respond to anyone who doesn't yet understand the importance of blameless discussions in which finger-pointing is nonexistent.

Keeping it blameless

A post-incident review must be *blameless,* which isn't the same as no accountability. Everyone makes mistakes, and the team must share responsibility for the decisions that led up to the incident. You are a team. You win together and you lose together. No one individual on your team should ever be used as the scapegoat for an outage.

Humans have an almost instinctive need to assign blame, and often that blame comes with a designation of being a "bad person" or a "bad engineer" — as if the person made the faulty decision out of malice. The dangers of a negative and blame-filled post-incident review are countless. When people feel as if they'll be punished — or fired — for telling the truth and highlighting their mistakes, they cover their tracks. Collaboration nosedives and much of the work you've done to transform your organization to a DevOps culture is lost.

Make post-incident reviews as positive as possible. Remember, you're looking for failings in the systems and processes you've established, even if a human was the one to discover the issue. If blame starts to seep into the conversation, leadership *must* step in and remind the group why a culture of learning is important and how blameless post-incident reviews fit into your attitude of embracing failure.

Reviewing the timeline

Earlier in this chapter, I step you through how to manage an incident as it's happening, but one of the things I point out is the importance of establishing a timeline. When you start your post-incident review, start with the timeline. Review what your engineers' first instincts were when facing the problem. What data did they seek out? Did your monitoring, alerting, and logging all give you the information you needed and expected? What was missing?

In addition, look at parallel work. Incidents aren't linear, clean events; instead, they're messy, and everyone scrambles to fix the issue as quickly as possible, which means that different people work on different things at the same time.

Figure 18-4 gives you an idea of a possible timeline. As you can see, Engineer 1 received an alert at 6:20 p.m. indicating something was wrong. A few minutes later, they realized that they weren't capable of handling the incident independently and escalated it to the second engineer on call, Engineer 3. At this point, Engineer 1 stepped back from technical contributions and instead acted as a communications chief and records the incident. At 6:34 p.m., Engineer 3 created a dedicated channel in chat for the incident. Engineer 4 quickly joined and was subsequently followed by Engineer 2. While Engineer 2 dug into a service he thought might be the issue, Engineer 3 and Engineer 4 worked together to review the logging and discover the issue. After they located the problem, Engineer 2 supported the efforts of Engineers 3 and 4 to bring the service back online. They resolved the incident at 7:01 p.m.

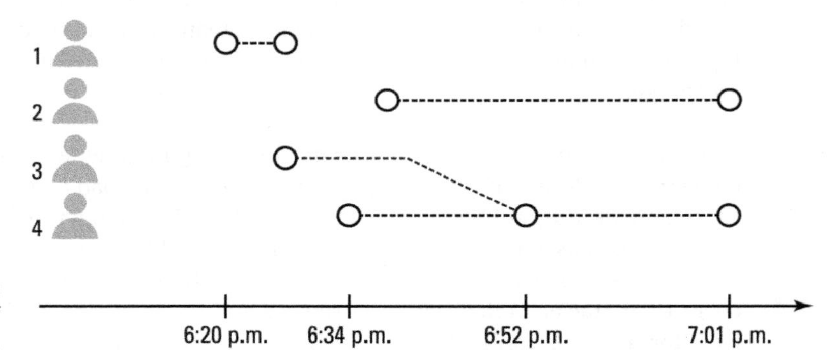

FIGURE 18-4: Timeline of an incident.

You don't have to have a perfect timeline or spend time drawing it. The point of this illustration is to see the parallel work and identify ways in which your team can collaborate and communicate more efficiently during an incident.

Asking tough questions

A post-incident review is most impactful when you can fully dig into the areas in which your team needs to improve — both technically and socially. Create space for people to share their perspectives and think through what could be better. Here are questions to help you get the conversation started:

- How did you discover this incident?
- Did alerting reveal the incident alerting or did someone manually stumble onto it?

- Did anyone notice the incident noticed in a timely manner?
- How long did it impact customers before the team was alerted?
- Did the service's telemetry provide the necessary information?
- What changes to monitoring, alerting, logging, and dashboards would help notify you about this particular problem faster in the future?
- Is the service stable moving forward?
- What work does the team need to do to reinforce the service's resiliency?
- What automated tests should you add to ensure that this particular incident won't occur again?
- Does someone need to write additional documentation?
- How can you help engineers on call quickly identify this type of problem?
- Did anyone experience an access limitation during the incident that prevented them from fixing the issue?
- What initial actions did you take in response to this incident?
- Which actions were a net-positive? Which attempts were a net-negative? What work had no impact at all?
- Did the incident impact any data? If data was lost, can you restore it?
- Do you need to notify any customers notified of collateral damage resulting from this incident?
- Did a deploy kick off this incident? If so, did the engineers experience any friction in rolling back the deploy or cherry-picking a previous release?
- How can you decrease the time involved in discovering and resolving the incident?
- How can you reduce the number of customers impacted if a similar incident occurs?
- Do you need to make changes to your development workflow, CI/CD pipeline, or release process to prevent future failure?
- Does anyone want to add anything?

Be sure to allow room for additional thoughts and random ideas that don't fit neatly into a prescribed list of questions. You're having a conversation, not an interrogation. Also, the review is as much a bonding opportunity for your team as it is a chance to uncover hidden gaps in your system.

Acknowledging hindsight bias

In a post-incident review, you have the gift of hindsight bias. You are reviewing past decisions with a fine-tooth comb while knowing the outcomes. The people who made decisions during the incident didn't have that benefit. They made tough decisions within the constraints in which they were forced to work.

Assume positive intent. Almost no one purposefully tries to sabotage their colleagues. Engineers take pride in their work, and everyone on your team is doing the best they can. With hindsight bias, you can easily overestimate the predictive ability of the people whose decisions you're criticizing. The truth is that solving hard problems is, well, hard. Really hard. Mistakes happen, so be kind to your past self and the past selves of others.

While you're at it, listen to dissenting views. People who disagree with the crowd can stumble on particularly interesting theories. Everyone comes to a situation with a different experiences, context, and viewpoints. That diversity is a gift that can help you better understand the intricacies of your socio-technical system.

Taking notes

Have one person in the meeting record the conversation and store the notes in a place everyone can access. You can take this note-taking idea one step further by recording the audio of the conversation, but only if you feel that everyone will still feel comfortable speaking openly and fearlessly. Taking notes of the meeting accomplishes several things. First, it ensures that anyone who couldn't make the review can still find out the details of what was discussed. Second, it provides new employees with insight into previous incidents as well as how the company responds to unplanned work. Finally, the notes give you evidence of a process that works when you're confronted with naysayers in the organization. If an executive wants to know why you're spending two to three engineering hours on a meeting to discuss something that's already fixed, you can educate them on how this work is a way to prevent unnecessary failure in the future and make your services more resilient.

At the end of the review, have one person compose a summary of the meeting for customers and internal stakeholders. When writing external messages, be sure to leave out any confidential information relating to the business, including the names of your engineers who don't want to be identified. As the folks at PagerDuty point out in their post-incident review documentation, avoid using the word *outage* unless it truly was a full-on site outage. *Incident* or *service degradation* gets the point across without making the situation seem worse than it was.

Making a plan

After you have a good idea of what went well, what went poorly, and what areas of your system need some work, make a plan to complete that engineering effort. Create the necessary tickets or free time for your engineers to reinforce the areas that will make your system more resilient and less brittle. You should prioritize this work, including by making space for it in the next sprint or week of work. Be sure to assign the work to specific people who can "own" the completion of it. Then follow up. After you've determined an estimated due date, make sure to loop back to see whether everything went well or more work is necessary.

5 Tooling Your DevOps Practice

IN THIS PART . . .

Discover how to modernize your software architecture by taking advantage of open source software and adopting new languages.

Manage distributed systems by designing decoupled microservices, standardizing APIs, and containerizing your applications.

Find out how to choose the best cloud provider and migrate your systems to the cloud.

» Licensing in OSS

» Running applications in containers

Chapter 19
Adopting New Tools

Continuous improvement and rapid iteration are fundamental to DevOps. That means your systems will constantly be changing and you'll need to adapt your technical approaches. New languages, frameworks, libraries, and tools are being developed all the time. Balancing maintenance and stability with adaptation and iteration can be difficult. You can make all these decisions with DevOps in mind, centering the customer and ensuring collaboration among your team.

Any time you integrate a piece of software — open source or commercial — into your system, you must consider the overall demands of the system and how each piece of software will communicate and interact with every other component. The best solution in the world is useless if you can't seamlessly integrate it with your existing system. Similarly, if a tool is difficult to use, problematic to extend, lacks documentation or at risk of being deprecated, you should hesitate to select it.

Third-party software (tools created by another person or company) must be flexible and resilient. Otherwise, they won't work well with the tools you already rely upon. In this chapter, I review what open source software (OSS) is, how you can benefit from integrating it into your system, and how to select languages in which to write new services.

Integrating with Open Source Software

Open source software provides individuals and companies with high-quality solutions to difficult-to-solve problems for low or no cost. The openness often implies that the tool is, well, open. You can go and view the code yourself as well as often clone the repository and use the tool as a foundation on which you add functionality. In contrast, "closed" software is typically proprietary and owned by a commercial enterprise. You cannot simply dig into the code that buttresses the software. You must trust the company to have developed a tool that is secure, dependable, resilient, and fault-tolerant. Each option has benefits and risks to your business, and often you have to decide based on each tool and offering. Open source software is not, as a rule, always better, and commercial tools don't always fulfill the benefits described in the sales call.

Before I talk about integrating open source software into your system, I need to define what "open source" actually means. Too often, people use the term to describe multiple aspects of the industry, which can lead to miscommunications and poor decisions.

Opening community innovation

The term *open computing* covers a wide variety of topics related to community innovation, but it's used interchangeably with *open source.* People in the industry have multiple points of view on this topic, so you'd be wise to seek dissenting points of view as you make decisions about the role of open computing and open source in your application.

Open standards

Since the Internet's inception, people have relied on open standards to make it function. Standard protocols are what allow the widespread network to communicate and function. These protocols include everything from HTTP (Hypertext Transfer Protocol Secure) to SMTP (Simple Mail Transfer Protocol) to TCP/IP (Transmission Control Protocol/Internet Protocol), all of which are used in billions of Internet information transfers every day. The industry relies on markup languages like XML (eXtensible Markup Language), YAML (YAML Ain't Markup Language), and JSON (JavaScript Object Notation) to serialize data in a (semi-) human readable way. Even programming languages have standards committees that make decisions on the best ways to implement features.

If these standards weren't open, industry innovation wouldn't be possible. The situation would be like a hundred road companies building an interstate without any plan for how to orient the roads, create connections, and develop uniform road materials.

Open architecture

Flexible architecture decisions are critical for a DevOps organization. Your technical system can grow and evolve, as can the engineers who maintain it. Open architecture describes the standard interfaces engineers use to connect independent components. Services Oriented Architecture (SOA) is an example of a design style that creates reusable and reconfigurable components that implement functionality. Application Program Interfaces (APIs) use a variety of standards such as REST (Representational State Transfer) or GraphQL to enable applications (or microservices) to interact.

Open source

Open source software (OSS) refers specifically to software released with the source code visible to anyone. You may copy, modify, and distribute the original work — all without royalties to the original creator. OSS has given the industry some of the best software currently available. Linux, Python, Eclipse, and Mozilla's Firefox are all examples of OSS. Open source software has formed the basis for many of the commercial products you use every day, including the operating system for your mobile phone.

Licensing open source

Licensing plays a key component in OSS. The term *free software* was defined by Richard Stallman of MIT in the 1980s as meeting four conditions, which he referred to as the four freedoms:

- Use
- Study
- Share
- Improve

Eric Raymond and Bruce Perens founded the Open Source Initiative (OSI) in 1998 (`https://opensource.org/history`) and determined the criteria of OSS. Proprietary software may be free, but that doesn't make it OSS. Unless you can view and modify its source code, a product is not open source. For a product to be considered OSS by the OSI, it must meet the following ten conditions:

- The license must allow anyone to sell or redistribute the software without royalty.
- The source code must be distributed along with the product.

- The license must permit modifications to the original code.
- The OSS license may permit restrictions to protect the integrity of the author's source code, such as requiring a different name for derived works.
- OSI prohibits discrimination against any person or group.
- The license must not restrict the use of the software for any particular purpose.
- Licensing is distributed and there is no need for additional licensing upon redistribution.
- The rights given through the license apply to anyone and do not depend on a product or redistribution vehicle.
- The license must not restrict any other software potentially distributed with the OSS.
- The license must be neutral to specific technologies, tools, or standards.

Hundreds of OSS licenses exist, each with its own unique spin on what's permitted. Be careful to check the licenses to ensure that you're in compliance.

Licensing your open source software or evaluating the licenses of OSS that you want to utilize doesn't necessarily require a lawyer. For example, the MIT License used in many products is rather short and readable. You can find the current MIT License at `https://opensource.org/licenses/MIT`. I recommend looking at some other common OSS licenses, such as GNU General Public Licenses (GPL) and Apache License, to get a feel for what you can expect when adopting OSS with different licenses.

Deciding on open source

Often the best option is to combine open source and proprietary software. At this point, open source has reached an adoption rate that has forced commercial software to ensure compatibility. OSS offers a number of benefits to companies that take the time to research and adopt it. Still, as with any engineering decision, you always have some gotchas to consider.

Benefits

Many of the benefits of open source software relate to its availability, cost, and general quality:

- **Low up-front costs:** OSS provides extreme monetary benefits for companies with low-to-zero costs upon initial adoption. The software must be maintained and integrated, so engineering hours will be required but the overall cost is typically a fraction of developing the tool in-house.
- **Quick acquisition:** Unlike some proprietary software that requires trials and pricing negotiations, OSS is often as simple as a quick download. (Okay, sometimes it's a slow download; I'm looking at you, JVM.) The other benefit to quick access is that developers can create minimum viable products (MVPs) without difficult-to-obtain manager approval. Their curiosity can drive their personal innovation and allow them to explore the opportunities of a product without any buy-in.
- **High-quality engineering:** With OSS, you conduct a peer review and integrate all contributions into the project but another contributing engineer performs a review. This community engagement makes for decisions that have been well thought out and evaluated. The community-led development ensures the involvement of a group of engineers who are both deeply knowledgeable about the source code and community oriented. The result is often robust communities of people who are willing to help others with questions as well as create documentation and tutorials.

Drawbacks

Potential drawbacks to consider relate to the engineering effort you need to make to integrate and maintain the software:

- **Lack of support:** One of the main draws to commercial software is the support provided. You can access robust documentation and help in implementing the software. Depending on your contract, you can access support employees dedicated to customer success. If this support is critical to your success, I recommend a choice beyond most OSS offerings.
- **Integration challenges:** You and your team will be solely responsible for integrating the OSS into your existing systems. This integration is often more complex than originally expected because of system surprises and limitations of legacy code. If by some chance you can deliver on time or ahead of schedule, yay! But never count on that one-in-a-million type of luck.

TIP

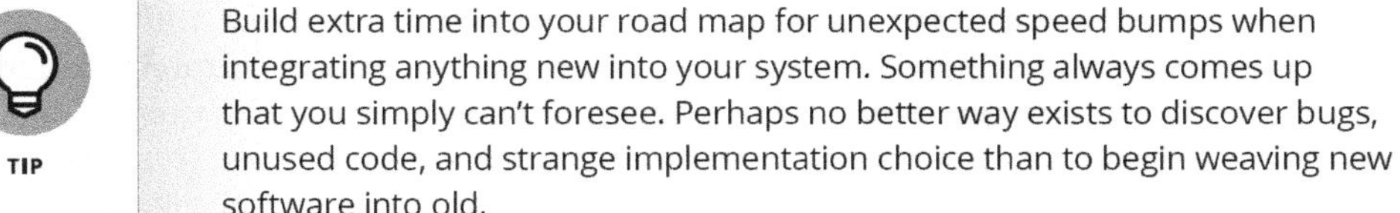

Build extra time into your road map for unexpected speed bumps when integrating anything new into your system. Something always comes up that you simply can't foresee. Perhaps no better way exists to discover bugs, unused code, and strange implementation choice than to begin weaving new software into old.

» **Maintenance:** Some open source solutions are immature. They're young and poorly adopted, which doesn't mean they're bad products but does limit the feature set and community around the product. An immature product is more difficult to maintain and requires a more focused engineering effort. If your business can sustain such efforts and the benefits outweigh the cost, go for it. But think through the long-term viability of products before you integrate them.

Transitioning to New Languages

Deciding to adopt a new language or framework is a common and sometimes horrifying prospect. Just as with spoken languages, programming languages share common structure. After you understand the basic parts of a software language, you can generally transfer that knowledge to another language. Most statements such as "Python is better than Java" have more to do with the engineer's comfort level than reality. Simply put, you develop faster and better in a language you know well.

Languages differ mainly on syntax and paradigms. But requirements such as special considerations and operating systems can also impact your decision about what language to adopt.

Although programming languages can have wildly different syntax, most languages allow for multiple paradigms. You can write JavaScript functionally, imperatively, or by using object-oriented programming techniques. You can write Go to be imperative or procedural. Python covers just about any flavor from compiled to interpreted. Python can be functional, object-oriented, iterative, or reflective. If you can think it up, Python is likely flexible enough to handle it.

A few reasons exist, however, to consider some languages over others, and they relate to the technical needs of the product and your engineering team (currently and in the future).

Compiling and interpreting languages

C, C++, C#, Erlang, Elm, Go, Haskell, Java, Rust — along with others — are all compiled languages. JavaScript, Ruby, and Python are interpreted languages. The main difference between compiled languages and interpreted languages is how the machine reads the program. People speak human (whatever version of human you happen to speak) but machines speak, well, numbers. People have to reduce the verbose nature of our language to ones and zeroes for the computer.

An interpreted language uses an interpreter (another program) to parse the instructions of the program and then executes it. An interpreted language requires no consideration of infrastructure beyond having the interpreter installed. On the other hand, compiled languages translate a program into the assembly language of the computer in which the program runs. The architecture of the computer must support the language into which the program has been compiled.

A compiled language typically performs faster because it uses the native language of the computer. Think about how much faster you can speak and comprehend your native language than your second or third language. No translation step is involved; you simply understand. It's the same for computers. In addition, compiled languages provide an opportunity for optimizations during compilation. An interpreted language is easier to implement and runs immediately. It needs no compilation stage after a change or update.

The compute power and tools you have at your disposal today make this distinction between compiled and interpreted languages much less important than in the previous decades. Although improved hardware has reduced key processing and resource allocation decisions to allow you to focus on other things, recognizing the differences between languages can give you a deeper understanding of the potential benefits and pitfalls of your decision to adopt a specific language.

Parallelizing and multithreading

When I first learned to write code, I was consistently confused about the difference between a language or system that is concurrent and one that is parallel. They seemed to be pretty much the same thing However, they are different, though the difference is rather pedantic. A concurrent system can support more than one action *in progress* at the same time, whereas a parallel system can support more than one action executing simultaneously.

In other words, a parallelized system is executing two separate commands at the same time. A concurrent system might appear to execute in parallel but instead assign both tasks to the same thread. Parallelism requires multiple processing units at the hardware level.

Multithreading is a related, but slightly different, process to parallelizing. With multithreading, the operating system (OS) executes multiple processes at the same time while sharing computing resources. The central processing unit (CPU) executes more than one command concurrently.

Languages that can improve your ability to powerfully parallelize your systems include:

- Ada
- C#
- Clojure
- Elixir
- Erlang
- Go
- Java
- Rust
- Scala

Programming functionally

Functional programming is a style that eliminates — or significantly reduces — mutable data by avoiding changing state. Within this paradigm, functions are idempotent (unchanged). The result of a function is dependent only on the arguments passed to the function and cannot be impacted by local or global state. If you want clean code, adopting functional programming practices is a solid start.

Whether you adopt a functional language or simply integrate the concepts into your code standards, you should understand three important concepts of functional programming, which the following sections explain.

Higher-order functions

These functions take other functions as parameters. This type of function allows for currying — a way of forcing functions to return new functions that accept arguments one at a time.

```
function doSomeMath(n, task) {
  return task(n);
}

function addOne(n) {
  return n + 1;
}
```

```
function subtractOne(n) {
  return n - 1;
}

doSomeMath(3,subtractOne);
```

Pure functions

A pure function has no side effects. No dependency exists between pure functions and one has no way of interfering with the other, so they're thread-safe and can be executed in parallel. The following code shows a pure function in contrast to an impure function. In the latter, a parameter must be accessed from outside the function, which is unacceptable in pure functions.

```
var pureFunction(a, b) {
  // returns the sum of two values passed into the function

  return a + b;
}

var impureFunction(a) {
  // b is not a parameter and therefore must
  // be accessed from outside the function

  return a + b;
}
```

Recursion

In functional programming, you accomplish iteration most often by using recursion. A recursive function can invoke itself. When you can use recursion, you should because it's considered to be more elegant and resistant to bugs. This code shows a simple recursive function that counts down from a specified number (and prints the numbers in the console), stopping at zero.

```
function subtract(n) {
  console.log(n)

  if (n === 0) {
   return 0;
  }
```

```
    else {
     return (subtract(n - 1));
   }
  }

  subtract(10);
```

You can adopt the preceding concepts in any language, but they are more common in functional languages or languages that better support functional paradigms. People consider Elm and Haskell to be purely functional languages, but you can write Java, Scala, Closure, and even JavaScript functionally.

Managing memory

Memory management is a way of allocating memory to specific functions. Applications require memory management to ensure that a running program has the resources available to provide any object or data structure the user demands. Memory management involves initial allocation and recycling, or garbage collection. The allocator assigns a memory block to the program, and when a block is no longer needed, the garbage collector makes it available.

In some languages, the programmer must manage the garbage collection process. In other languages, the approach is automated. C#, Go, Java, JavaScript, Ruby, and Python take care of garbage collection for you, whereas languages like Rust and C require manual memory management.

Choosing languages wisely

The point of this section isn't for you to memorize the pros and cons of every language. Instead, it's to drive home that you have a wide variety of languages from which to choose. Some are prescriptive in their approach to programming; others are endlessly flexible and easily manipulated. Some are specialized for a specific use case. For example, R is best known for statistical computing. Java is known for enterprise applications. Python is widely used by data scientists and web developers alike. Go is especially useful for high-performance and runtime efficiency. Swift is designed specifically for Apple devices.

Beyond language characteristics, you need to keep five questions in mind when choosing your next language, as the following sections explain.

What is the quality of the language community?

Choose a language whose community aligns with your company culture. A small subset of languages is known to have stellar communities. The Ruby community, for example, is incredibly welcoming, diverse, and concerned with taking less experienced and junior engineers to higher levels. Ruby is easy to learn simply because of the number of engineers who are ready to pitch in and help you get the hang of things.

A healthy language community provides a number of benefits:

- Widely available (and usually free) mentoring
- Blogs and tutorials to help developers get started
- Plenty of answers to questions on programming forums

REMEMBER

Ruby is the language in which I learned to program, and I will be forever grateful to the groups of engineers who helped me get my start. They provided a safe environment in which I could ask questions — even dumb ones. Help others without judgment and without expectation of repayment is an important aspect of the community. That is how you grow a tech community that is healthy and enduring — for everyone.

How many developers know the language?

Select a language that will allow you to recruit from a large pool of candidates. Some languages are so widely adopted that finding a talented engineer is a relatively easy feat. Others are either so old, deprecated, or specialized that the engineers who can maintain legacy systems and add new code are incredibly rare (and typically well paid). Perl and ColdFusion don't have a ton of language specialists left. On the other hand, you can find an almost endless list of brilliant JavaScript engineers. You don't need to hire experts in a particular new technology. Yes, having an anchor on your team who knows a great deal about something super specific would be extremely helpful. However, the need to hire people with curiosity and a passion for finding new engineering solutions rises above all other considerations.

What frameworks and libraries are available?

Frameworks and libraries can make or break your project. Seriously. Using a language that provides particular libraries that allow your engineers to call code out of the box can shave months off your project. A framework is a lot like a scaffolded application that's ready out of the box. It's a skeleton onto which you can add functionality. In many ways, a framework defines design paradigms. As a result, you most often find frameworks in flexible languages. Python has Django; Ruby has Rails; JavaScript has React, Vue.js, Node.js, Angular, Polymer, Backbone.js, Ember.js, and so on. A library does the work in creating complicated but often used algorithms for you. Mathematics and physics libraries, for example, allow you to call complicated functionality without doing the algebra yourself.

What are the specific requirements of the project?

As you saw in the previous section, languages come in all sorts of shapes and sizes. Some are prescriptive whereas others will contort themselves to allow you to do whatever janky fix you're working on. No one language is better than the other, but they can be better suited for specific projects.

If your project has a mobile component, how you approach your language choices will need to take into account mobile app development for Android and iOS. You also need to take into account the physical environment within which your project or team will work. If much of your infrastructure is based in Azure or Microsoft servers, a Microsoft solution will likely be more easily implemented. Finally, keep in mind the scalability requirements of the project. Some languages scale more simply than others.

What is the comfort and knowledge of your current team?

Ideally, you choose a language with which your team is at least vaguely familiar, or one that they can learn quickly.

Tossing a brand-new language or framework at your team without any foundation of knowledge will fill them with imposter syndrome and some sense of dread. What happens when they're found out as frauds who don't actually know what they're doing? Of course, they're not frauds. You hired talented engineers who are capable of taking on challenges. But it is the emotional response you'll likely receive.

If a project truly calls for something completely new, try to find a connection for your engineers. Perhaps you're moving your configuration management to Chef — a tool with which no one on your team is familiar. But perhaps one of your engineers was formerly a Ruby developer. Simply having that one piece of familiarity will go a long way toward taking the team to a higher level quickly.

» Designing APIs

» Running applications in containers

Chapter 20
Managing Distributed Systems

A distributed system is simply a collection of components networked across multiple computers. The components are independent (or at least should be), can fail without impacting other services, and work concurrently. Services communicate with each other through messaging formatted for a particular protocol (like hypertext transfer protocol, or HTTP).

Decades ago, the server that hosted a company's application often lived in a closet at the office. (A few of you might still have old remnants of hardware in office closets.) Now, the majority of companies are beginning to take advantage of pay-as-you-go cloud hosting. In large part, this move to cloud hosting is happening because running applications at scale requires efficient use of infrastructure. The costs of underutilizing hardware add up quickly.

Distributed systems have become the norm, mainly because of cloud services. Multitenancy allows multiple customers to take advantage of shared resources, which keeps costs low by maximizing the use of those resources. If you use a cloud provider like Azure or AWS, the components of your system run on machines spread across a particular region (or regions). Most of the time, the cloud provider doesn't even know *which* machines your application runs on.

I talk more about moving to cloud platforms and infrastructure in Chapter 21. For now, I focus on the two concepts that accompany this transition to distributed systems: microservices and containers. Microservices are a style of architecture that separates logic into loosely coupled services. In theory, the modularity makes the application more resilient and easier to maintain than a monolithic application. Containers enable engineering teams to package applications with dependencies and provide an isolated, ephemeral environment.

In this chapter, I dig into transitioning from a monolithic architecture to microservices, explain how APIs enable distributed systems, and discuss working with containerized infrastructure.

Working with Monoliths and Microservices

Whatever language and tooling you choose, you must merge all the pieces into a working system. The two common architectural structures in modern applications are monoliths and microservices (with microservices leading as the popular choice for high-performing teams).

If you're wondering whether microservices are just service-oriented architecture (SOA), you're mostly right. SOA architecture has a few key characteristics:

- Components or units of functionality logically manage business functions.
- Every unit is self-contained.
- Users don't need to know how a component works, only how to interact with it.
- Other services can exist within a unit, but components are loosely coupled.

Microservices are a modern implementation of SOA. Although no industry standard exists for what constitutes a microservice, you can start with a few basic principles.

Just as with SOA, microservice architecture is loosely coupled. Units of logic — services — are reasonably separated. You can update and deploy services independently. Microservices are small — it's in the name! — and accomplish one piece of business functionality. You can write services in different languages and support them with different infrastructure. The units communicate with each other via tech-agnostic interfaces and protocols such as APIs and HTTPS requests. The modular nature of microservices makes an application easier to read, comprehend, troubleshoot, and maintain.

This separation of logic improves service ownership across your engineering team. It also allows teams to independently choose language and tooling while still applying DevOps principles across your organization — improving autonomy and enabling collaboration. Although every single piece of logic doesn't need to be abstracted into a microservice because of unnecessary complexity, breaking out logic into smaller components will benefit your team and enable you to more smoothly adopt continuous integration and continuous delivery (CI/CD).

Choosing a monolithic architecture first

My personal preference is to start a brand new application with a monolithic architecture. If you're a startup or just getting started on a minimum viable product (MVP; see Chapter 7 for details about MVPs), don't bother overthinking your architecture. Yes, making some key architectural decisions with growth in mind is important, but I'd argue worrying about how to dynamically scale an application while you have 20 users is a poor use of time.

Figure 20-1 visualizes a monolithic application. A user interface (UI) communicates with business logic. The functions that make up that business logic have access to a data layer that finally communicates directly with the database. Data flows up and down the stack.

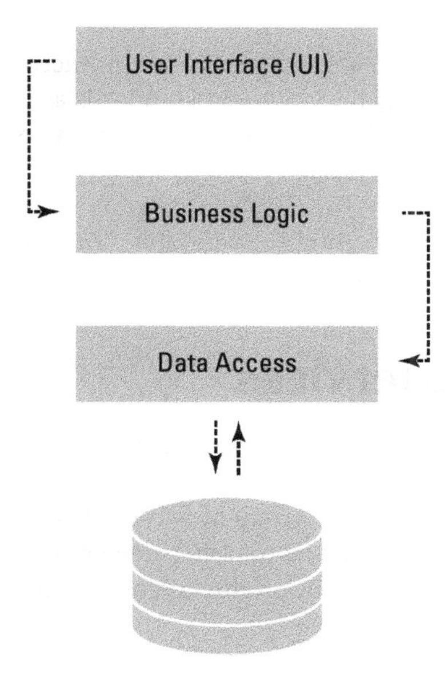

FIGURE 20-1: Monolithic architecture.

At some point, you start to notice friction along the software delivery life cycle. Developers step over each other when trying to fix bugs or add functionality. A bug in user logic will cause a service disruption when users attempt to buy a product. As these points of irritation pop up, consider slowly decoupling logic into microservices. You will pull functionality out from the monolith into smaller components.

WARNING

Be sure to delete unused code. Failure to delete unused code is the most common mistake I see when people begin to strangle a monolith and adopt microservices. Don't be afraid to delete code. You use source control, and you have access to previous commits and build. Nothing in a codebase acts more like a landmine than unused code. A single duplicate name added months or years later that accidentally calls old logic can quickly cascade into a massive failure.

As you slowly pull logic apart into neat and tiny components, at some stage you'll find that you have a planetlike monolith with microservice moons rotating around it. I encourage you to sit in that stage until you know that your engineering team can manage a fully decoupled system.

If you *think* you have microservices but actually have what I lovingly refer to as "macroservices," you will find yourself in a much more tangled situation that is difficult to undo. Unlike microservices, macroservices are linked together by tenuous and nearly impossible-to-detect ways.

When you decide you're willing and able to go full-in on microservices, it's time to get serious about code quality and development standards as well as ensure that you have clear standards regarding API design and versioning so that services can communicate with one another seamlessly.

REMEMBER

A well architected monolith is much preferable to sloppily developed microservices.

Evolving to microservices

Beyond the decoupling of logic, well-architected microservices offer large engineering organizations the capability to split teams by components in which each group of engineers "owns" a service (or group of services) from ideation to production. This type of architecture enables parallel development of features by multiple teams. (It also requires a team of product owners or project managers to appropriately divide work.)

In contrast to a monolith, microservices can interact freely with each other, and those services will pass information around until data needs to be saved or retrieved from the database. Microservices involve a much more free-form architecture, as you can see in Figure 20-2.

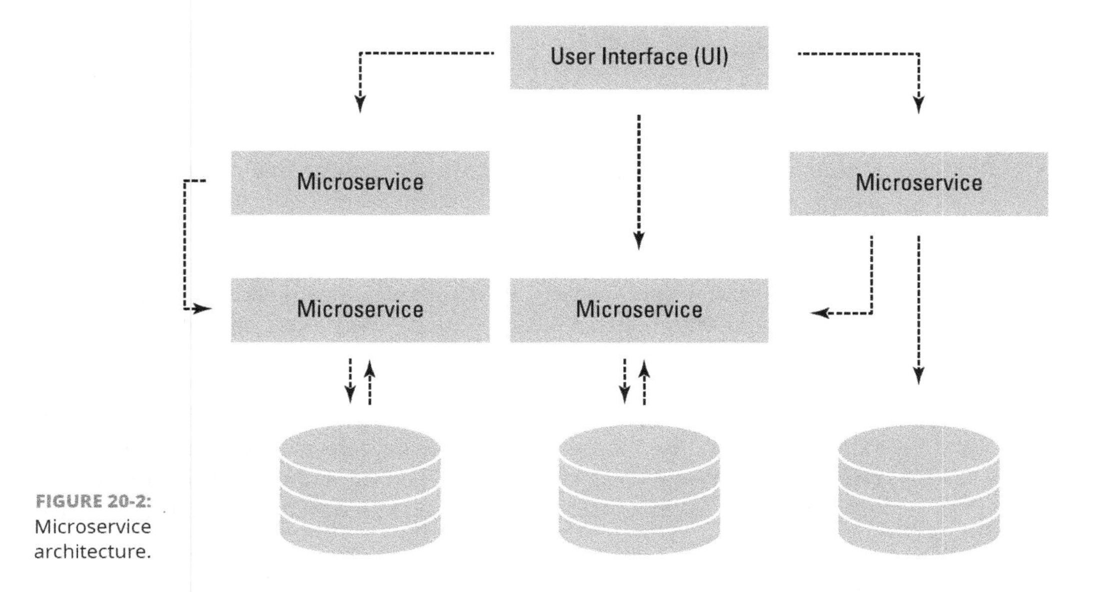

FIGURE 20-2: Microservice architecture.

A team can develop, test, and release each component through the team's unique CI/CD pipeline. Every team can own any technical debt they choose to take on throughout the process. If a particular feature is best designed in Go whereas another is ideally implemented in Python, both can exist in their independent state and communicate through a language-agnostic protocol. In addition to (potentially) not sharing a language, services won't share databases or hardware.

From an operations perspective, microservices simplify deployments. Because microservices are typically smaller in nature, with fewer lines of code, than a monolith, you can more easily deploy small changes frequently, thus eliminating a common challenge in adopting continuous delivery or continuous deployment. Perhaps most important, microservices enable refined and targeted scaling. Instead of allocating resources to the entire system, you can pinpoint specific services that have increased demand and allocate additional resources for that component.

You will find repeat logic throughout your application. Do not duplicate code in different services. Instead, create libraries containing shared logic that any service can access. Be sure to adopt a message-queuing solution that uses a format like JSON to appropriately notify services of changes in a nonsynchronous manner. You can design message queues to persist, thereby eliminating data loss in transit.

Any team that attempts to move from a monolith to a microservice architecture will experience challenges. Triaging those issues is a small price to pay for the potential gains you can achieve. When you design services in a way that creates conflict between teams, apply DevOps processes to communicate what each team needs and adapt the independent services to work together.

Designing Great APIs

An application programming interface (API) is a way to expose only the specific objects and actions required. An API can be anything that a human (or computer) uses to interact with a software application. When I talk about APIs and services, however, I refer to RESTful (representational state transfer) APIs. The RESTful approach takes advantage of HTTP (HyperText Transfer Protocol) requests that allows applications and services to communicate with each other. HTTP has four actions (sometimes referred to as verbs): `GET`, `PUT`, `POST`, and `DELETE`.

TECHNICAL STUFF

Before REST, the default API format was simple object access protocol (SOAP). The advantage of REST is that it uses less bandwidth, making it preferable for transactions going over the Internet. APIs allow developers to expose their services to other developers and applications, which permits a system filled with diverse service designs to act as a whole. An API receives requests and sends responses.

What's in an API

In its most basic design, each API request requires two pieces of information: a noun and a verb. In other words, what thing are you interested in and what do you want to do to that thing? Specific identifiers may be required for certain actions.

The API then sends a response to the requester with an update as to what happened.

For example, say that a new user fills out the Sign Up form on your website. That form will likely collect the information inputted by the user and utilize an API endpoint that will pass the information to a service, which in turn will parse and validate the information. If everything checks out, the user is eventually saved to the database by the appropriate service.

The API endpoint for this request might be

```
POST /users
```

After the user has been saved to the database, the user is assigned an ID, which identifies that user as unique in the database. If you wanted to request that user, and that user's ID is 34, the endpoint might look like this:

```
GET /users/34
```

In this example, the HTTP action `PUT` would edit the user and `DELETE` would — you guessed it! — delete the user from the database.

Focusing on consistent design

API design becomes critical to organizations using widespread microservice architecture. Part of your internal structure should ensure that APIs are the *only* way services may interact. Shared memory, direct access to data, or direct linking will muddy your organization's processes and make bugs extremely difficult to find. APIs should serve as your system's exclusive communication channel, and you should design every service to use API access.

The suggestions in the following sections are far from exhaustive, but they describe with good practices to implement when designing APIs.

Using nouns

Create your endpoints to use nouns, not verbs. `GET /users` is preferable to `GET /getAllUsers`. Looking at it with the HTTP action should partially clarify why. `GET /users/34` is preferable to `GET /getUser/34`. Keep it clean. Simple, predictable patterns keep bugs at a minimum and ease the way for developers to design services for integration with your APIs.

Determining verbs

Be sure to use the correct HTTP verbs for the actions requested. `GET` fetches a particular object or a group of objects. `POST` creates an object or a collection of objects. `PUT` updates (or edits) an existing object or collection of objects. `DELETE` deletes an existing object or collection of objects.

Pluralizing endpoints

Decide on pluralization. API endpoints can use pluralized nouns for all requests or use singular nouns if appropriate. For example, if you use singular *and* plural, the endpoint to get all users is `GET /users` and the endpoint to get a single user is `GET /user/34`. My personal preference is to maintain consistency and use pluralized nouns for everything. To make sense of the idea in words, I think of it as `GET from USERS user 34` (which, from a database perspective, makes sense).

Adding parameters

Use extra parameters. You can pass as many parameters to the endpoint as you like. If you need to set up an API to fetch a user by name instead of ID, design it to look like this: `GET /users?name='emily'` rather than `GET /getUserByName`. The former keeps the design consistent and limits how many one-off API endpoints developers have to memorize or locate in documentation.

Responding with codes

Respond with appropriate HTTP codes. API design doesn't stop at the request. Developers must set up response codes to let the user or service know what happened after the request was received. Almost everyone has encountered a 404 page on a website. The 404 message is the response code for NOT FOUND, but you have dozens of options to include. Table 20-1 shows common response codes.

TABLE 20-1 Using HTTP Response Codes

Code Category	Code	What It Means
200 - It's All Good		
The 200 codes mean everything went as expected. But you can include extra information for specific responses.	200 OK	The most common HTTP response code. Everything was successful.
	201 CREATED	The POST request was successful. A new resource was created.
	202 ACCEPTED	The request was received but no action was taken.
300 - Please Come This Way		
Responses utilizing 300 codes are redirects.	301 MOVED PERMA-NENTLY	The resource requested has been changed. This is typically accompanied by a redirect URL.
	302 FOUND	The resource has been changed temporarily. Whereas 301 is permanent, 302 is not.
400 - User Error		
Any response code in the 400s is a client error. In other words, the user made a mistake.	400 BAD REQUEST	The server (or endpoint) couldn't understand the request. This is typically seen if incorrect syntax was used or as a default when what went wrong isn't clear.
	401 UNAUTHOR-IZED	Authentication is required. The client is not signed in.
	403 FORBIDDEN	The client does not have the correct authoriza-tion credentials to access the response. This error differs from 401 in that the client's identity is known.
	404 NOT FOUND	The requested resource cannot be located.

Code Category	Code	What It Means
500 - It's Me, Not You		
Finally, any response code above 500 refers to server errors.	500 INTERNAL SERVER ERROR	The server can't figure out what to do and needs to try again.
	501 NOT IMPLEMENTED	The server can't fulfill the request.
	502 BAD GATEWAY	The server received an invalid response when acting as a gateway to process the request.
	503 SERVICE UNAVAILABLE	The request can't be processed. Typically, the server is down and the request needs to be reattempted.

Versioning Your API

You should prefix the API version before the endpoint. For example, to get all users, use `GET v1/users`. You can increment subsequent versions however you see fit, although `v1, v2, v3` is simple and straightforward. But prefixing the version number ensures that a version is sent, which isn't guaranteed if it's sent as a parameter. This approach eliminates strange bugs when everything in a request looks good but the versioning is off. I strongly recommend that you account for backward compatibility as your APIs evolve and new versions are released.

Paginating responses

Take advantage of pagination to avoid returning overwhelming amounts of data or potentially bringing down the service. Be sure to set a default limit and offset that's applied to responses if none is supplied in the request. For example, the first page would return `GET /users?limit=25` — that is, the first 25 users (users with ID 0 through 24). The next page would respond with the data from `GET /users?offset=25&limit=25` and deliver the next 25 users (users with ID 25 through 49). `offset`, in this case, simply tells the service to skip the first 25 users when requesting the information from the database. You can increase the offset with each paginated request.

Formatting data

Choose a supported format. Most modern applications prefer information to be formatted as JSON in requests and responses. JSON uses fewer characters and is more readable than XML, though many applications still use the more verbose XML.

Communicating errors

Add error messages to give the user extra information, tailored to your service. Think beyond the response code. What does the user need to know? Examples include `RESPONSE 200 OK - The user was saved!` or `RESPONSE 400 BAD REQUEST - Missing required field: FIRST NAME.`

Containers: Much More than Virtual Machines

Containers are instances of a runtime object defined by an image. They are lightweight environments in which you can run your application. An image and a container are related but different concepts, and understanding the distinction is fundamental to deciding to containerize your application. An image is an immutable snapshot of a container. You can't change or update the snapshot. An image will produce a container if run using the appropriate command. Images are stored in a registry and ideally layered to save disk space.

Image layers are immutable instructions that allow a container to be created using references to shared information. For example, imagine building two containers that are identical up until the last two lines of instructions. Instead of building two containers from scratch, you can use layers to enable you to reference layer caches and rebuild only the last two layers.

```
docker run [OPTIONS] IMAGE [COMMAND]
```

Containers have isolated CPU, memory, and network resources while sharing the operating system kernel. They hold source code, system tools, and libraries. They differ in key ways from virtual machines (VMs), but you can think of containers as lightweight iterations of VMs.

Although containers have been around since the late 1970s, the technology wasn't mature enough to run applications in production until Docker debuted its platform in 2013. A modern container is a self-sufficient execution environment and repository for everything your application needs to run.

Understanding containers and images

Shipping containers are an often used but problematic metaphor for containers. I like to use Harry Potter instead. (Yes, I mean the wizard with the lightning-bolt scar on his forehead. Potter fans know that the scar is actually the motion of casting the Avada Kedavra curse.) I don't want to talk about Harry, exactly, but rather a concept that J.K. Rowling created for the wizarding world of Harry Potter: the pensieve.

In Harry Potter's world, a *pensieve* is a large, shallow bowl in which memories are re-created in a way that is faithful to the original environment — down to every detail — and can be experienced by anyone in precisely the way it originally happened. A memory is taken from storage and put into the pensieve. When a wizard or witch puts their face into the pensieve, they are thrust into the memory as if they were physically living it.

You can liken a container image to the memory that is stored without degradation until it's ready to be experienced by another wizard or witch through the pensieve. The container would be the reliving experience — an instance of that memory.

Deploying microservices to containers

Microservices are the ideal architecture to take advantage of containerized infrastructure. Because components are independent of other application functionality, they can be released and hosted individually. Microservices communicate via APIs, which means that different services can be released on different containers.

Figure 20-3 depicts microservice applications. Each application has many services that, when sewn together, make up the entire application's functionality. One service can be focused on users whereas another can be focused on order functionality. You have truly countless ways to divide an application's functionality into microservices.

After you divide an application into pieces that can be containerized, you can create immutable Docker images — those memories stored in the imaginary pensieve — which capture everything needed to run a service. Figure 20-4 depicts the images created for each service of an application, ready to be deployed independently to containers.

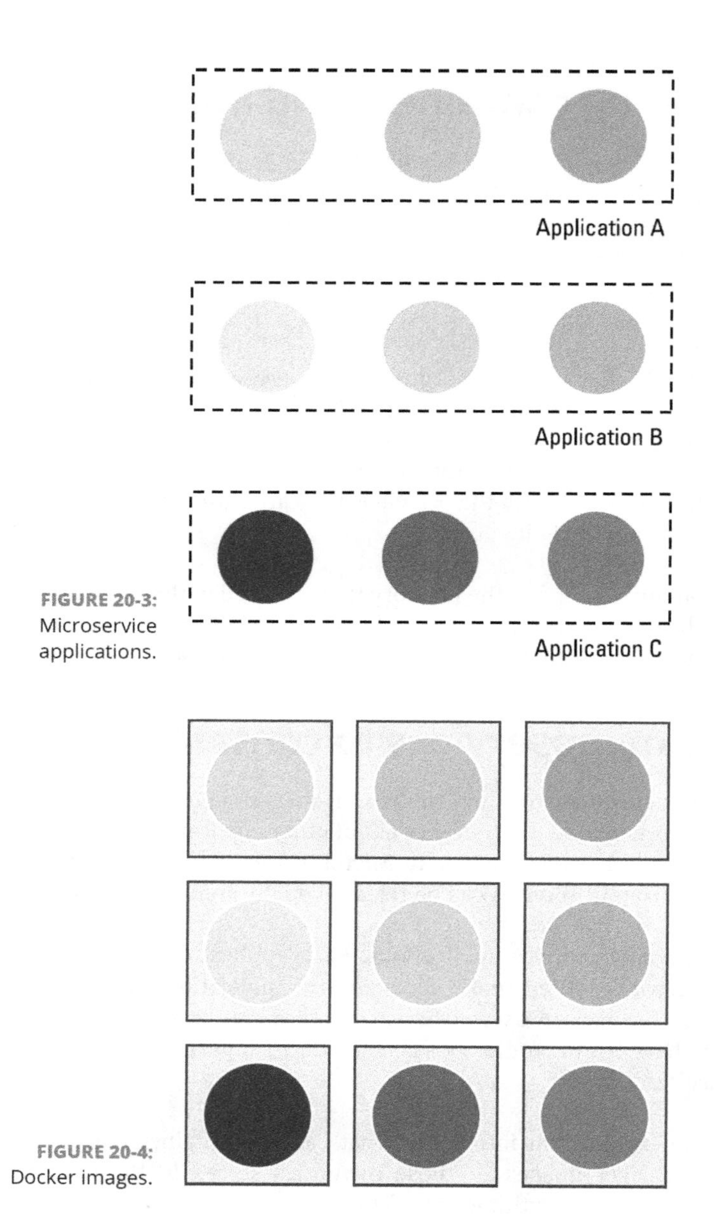

FIGURE 20-3: Microservice applications.

FIGURE 20-4: Docker images.

And finally, Figure 20-5 depicts running unrelated services in the same container. You don't have to group services in a particular way before releasing them. Containers don't care what you run on them, and microservices don't need to be colocated with any other logic. You can mix and match to maximize your use of resources, as well as group containers into clusters, as shown in the figure.

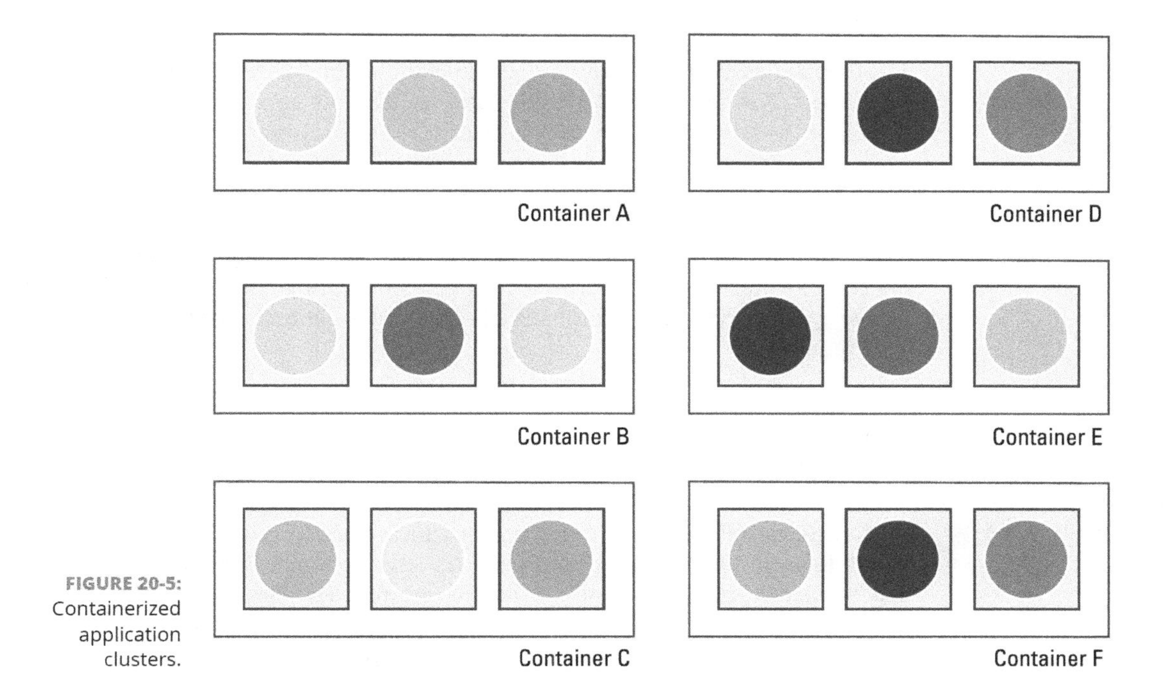

FIGURE 20-5: Containerized application clusters.

Containers enable distributed applications, but large clusters require orchestration to create, manage, and update containers across multiple hosts.

Comparing orchestrators: Harmonize the hive

Orchestrators help you manage sets of containers for applications running in production on multiple containers or using a microservice architecture. Getting visibility into complex systems through monitoring and telemetry for scaling is difficult, and an orchestrator can help you better manage your distributed system.

An orchestrator is essentially a manager that you can use to automatically scale (add additional resources) your cluster with multiple instances of each image, instantiate new containers, suspend or kill instances when required, and control each container's access to resources such as storage and secrets.

Many container orchestrators provide the features you need to run an application in production but refer to those features with different names. Smaller teams may prefer an open source solution to build from, whereas larger companies will likely prefer enterprise solutions that emphasize scalability.

The following sections describe five popular container orchestration and management tools. The tools described here comprise far from an exhaustive list but they highlight the largest communities and most mature solutions.

Kubernetes: The popular kid in class

Originally created by Google as a successor to Borg, Kubernetes — K8s or Kube, for short — was donated to the Cloud Native Computing Foundation in 2015 and is now open source. Its popularity as a Docker container orchestrator has exploded over the last few years.

You can use Kubernetes to manage containerized applications as well as automate deployments. Kubernetes simplifies the orchestration of containers across multiple hosts by managing the scale and health of nodes.

By sorting containers into groups referred to as "pods," Kubernetes streamlines scheduling workloads. It integrates well with other open source projects and enables you to quickly customize your infrastructure management. This customization includes maximizing resources, controlling deployments, and enabling your applications to self-heal through autoplacement, autorestart, and autoreplication.

Azure Kubernetes Service (AKS): Kubernetes and more!

Azure Kubernetes Service (AKS) is a managed Kubernetes orchestrator. It simplifies cluster management, deployment, and operations, but above all else, it simplifies the effort required to deploy Kubernetes clusters in Azure. AKS manages health monitoring and maintenance for you and helps you easily configure more complex integrations with Azure.

OpenShift

OpenShift is Red Hat's enterprise container application platform. Built on Kubernetes, OpenShift added features to enable rapid application development, easy deployment, and life cycle maintenance. It leverages automation and dynamically provisions storage. OpenShift is ideal for teams looking for enterprise-specific features and multitenancy.

Docker Swarm: More than a hive

Docker Swarm is the native clustering and scheduling tool for Docker containers. It uses the Docker CLI to deploy and manage containers while clustering nodes, allowing users to treat nodes as a single system.

Users create a primary manager instance and multiple replicas. This redundancy ensures continued uptime in case of failure. Manager and worker nodes can be deployed at runtime. It's a fast and scalable orchestrator. Swarm has been successfully scaled up to 30,000 containers.

Swarm is included in Docker Engine and, unlike other solutions, doesn't require initial setup and installation. Swarm enforces TLS authentication and encryption between nodes and allows rolling updates so that nodes can be upgraded incrementally.

Amazon ECS

Like other orchestrators, Amazon Elastic Container Service (ECS) is reliable, flexible, and extensible. It simplifies running Docker containers across Amazon Elastic Cloud Compute (EC2).

ECS is compatible with a serverless architecture, and you can use the built-in scheduler to trigger container deployment based on resource availability and demand. ECS is capable of scaling clusters to more than 10,000 containers, which can be created and destroyed within seconds.

Amazon ECS is ideal for small teams who rely heavily on Amazon and don't have the resources to manage bespoke orchestration and infrastructure.

Configuring containers

Although much of what you need to run a containerized application in production comes out of the box with Docker and popular orchestrators, you'll likely still need to configure a few things specific to your application. All configuration logic lives in a Dockerfile. If you have experience with Chef, Ansible, Puppet, or other configuration-management tools, a Dockerfile is the same thing that those organizations refer to as a cookbook, playbook, or manifest. It's a list of instructions for creating a container.

Orchestrators usually execute containers in groups — what Kubernetes calls "pods" — and supports additional configuration specified in a configuration file. At this execution, you can specify CPU and memory limits for each container. The benefit of resource configurations is that your orchestrator's scheduler can make more informed decisions about node placement. You can use namespaces to isolate configurations. The scheduler specifies CPU in units of cores and memory in units of bytes.

Nodes have a maximum capacity for CPU and memory they can allocate to containers. The scheduler ensures that the total of all resource requests is less than the capacity of the node, thereby eliminating resource shortages.

Monitoring containers: Keeping them alive until you kill them

Infrastructure and its ecosystem have evolved drastically over the last few years. A number of fundamental changes to application architecture and infrastructure have come about in the form of microservices and containers. Many monitoring tools and techniques are no longer relevant. Instead, engineers need solutions that can adapt to the short-lived and isolated nature of containers and application services.

Embracing increased complexity

The benefits of containers are flexibility, scalability, and portability, but these benefits come at a cost. Running containerized applications in production is significantly more complex because it involves more moving parts. Despite the downsides, the draw of a monolithic application is that everything is in one place. Your brain can keep track of what goes where.

Unfortunately, you don't get to have everything in one place in a containerized microservice application. Instead, the logic of your application is divided into hundreds of services running on hundreds (if not thousands) of containers. Your brain simply can't keep up. Here are a few key considerations to evaluate and address when adopting containers:

- **Containers are temporary.** You can create and destroy containers within seconds. The life span of a container is brief, sometimes only a few hours.
- **Containers are immutable.** Containers can't be updated. After an image is built, it can never be changed. Instead, a new image must *replace* it.
- **Containers are scalable.** The scalability of containers is an enormous advantage, but it also drastically increases the number of machines in your environment.
- **Containers lack persistent storage.** Unlike VMs or bare-metal servers, application data can't be stored directly in a container.
- **Containers require monitoring.** The performance and security of containers requires management through the use of an orchestrator or monitoring tool.

Container life cycle

Containers are short lived, and Docker provides basic commands to control the state of a container: create, run, pause, start, stop, restart, kill, and destroy. The life cycle of a container typically includes five states: defined, tested, built, deployed, and destroyed.

At the start of the life cycle, the container is defined via a Dockerfile that includes runtime, frameworks, and application components. Next, the source code is pushed through a CI system to be tested. The container is built and exposed to the orchestration system, where it is replicated and distributed throughout the cluster. Finally, because containers can never be patched, a container is destroyed and replaced.

Containers run in isolation and can be fairly opaque. With containerization, teams need to monitor services, hosts, and containers. Runtime visibility should track inputs, outputs, resource usage, and network stats.

Ideally, your runtime monitoring tool will be a small kernel module that can access the entire container ecosystem, which allows it to spot potential issues before they escalate out of control.

AVOIDING CONTAINER BLOAT

Runtime container bloat stems from the inefficient use of containers and degrades performance and scalability. Container monitoring agents need to be able to ingest service response times and network utilization as well as infrastructure and application metrics — without extra overhead within each container. You have two ways of implementing monitoring in containers.

The first is a sidecar container that utilizes groups of containers like Kubernetes pods — containers that share a namespace — and attach a monitoring agent within each pod. It's easy to set up, but resource consumption is high, and this approach creates another attack vector for a potential security threat.

The other way to implement container monitoring is to use transparent instrumentation, which utilizes a monitoring agent per host. Although this approach requires a privileged container and kernel module, it drastically reduces the monitoring agent's resource consumption. Transparent instrumentation requires a more complex setup but is well worth the effort because it allows you to collect ample amounts of data with little overhead.

Securing containers: These boxes need a lock

Software security is a never-ending and often uphill battle. New security measures and patches are released just in time for the next vulnerability to be exposed.

Securing containers in production can be especially difficult. Simply put, it involves more moving parts. You must secure every piece of the container ecosystem:

- Host OS
- Container runtime
- Orchestrator
- Container registry and images
- Application APIs and microservices

In addition, containers are relatively immature and evolving at breakneck speeds. Regular releases introduce change and require new security considerations.

Container contents are partially isolated from the host system but use kernel features, which makes using containers much more efficient than traditional using VMs. VMs thoroughly isolate processes and applications. In addition, containers share resources and can be created or destroyed instantly. But the lightweight, ephemeral nature of containers introduces new security vulnerabilities.

Securing secrets

Secrets are objects that contain sensitive information such as a username, password, token, key, or SSL certificate. This type of data should never be stored unencrypted in a Dockerfile or source code.

Docker secrets are encrypted during transit and are accessible only by services given explicit permission. Putting the data into a secret object reduces the risk of exposure. Secrets are typically accessed by a pod through a volume.

Potential vulnerabilities

VMs employ a hypervisor — a layer of abstraction that sits between the VM and the host. Containers eschew a hypervisor and instead act as the direct intermediary.

Containers are lightweight because they contain less information than a traditional VM, which is great for efficiency but requires additional security considerations. Here are a few areas you'll need to address in your container security strategy:

- **Container images.** You should secure images and registries. Containers should include only essential services.
- **Open source components.** It's important to have visibility into containers that include open source software. Regularly scan your images for open source vulnerabilities.
- **Shared kernel architecture.** By design, containers run on the host kernel. This shared resource makes containers extremely efficient, but it exposes containers to kernel vulnerabilities. Make sure your host and Docker configurations are secure.
- **Container breakout.** This situation occurs when a container has circumvented isolation checks and can access secrets or upgrade privileges. If a kernel vulnerability exists, containers can access the host.
- **Privilege escalation.** If a container mounts a host filesystem or Docker socket, the container can escalate its privileges. Limit default container privileges and limit Docker daemon user access.

WARNING

Don't store SSH servers inside images and don't host source code from containers.

» Choosing the right type

» Evaluating cloud providers

Chapter 21
Migrating to the Cloud

The cloud isn't some empyreal plane where bits and bytes are transferred through the atmosphere. It's just a fancy way of saying "someone else's servers."

"The cloud" refers to vendor delivery of cloud computing services. These services include physical infrastructure such as servers, data storage, networking, software services, deployments, monitoring, and more. Cloud services allow for faster innovation, flexible resources, and economies of scale. You typically pay only for cloud services you use, which helps you lower your operating costs, run your infrastructure more efficiently, and scale as your business needs change.

In this chapter, you find out how to apply DevOps to the cloud, benefit from various cloud services, understand the differences between private and public clouds, and choose the cloud provider that's best for you.

Automating DevOps in the Cloud

Marrying the cloud with your DevOps practice can accelerate the work you've already accomplished. When used together, both DevOps and the cloud can drive your company's digital transformation. Throughout this book, I emphasize the priorities of DevOps: people, process, and technology. The cloud — along with other tooling — falls squarely into the technical part of your DevOps implementation.

Cloud computing enables automation for your developers and operations folks in a way that simply isn't possible when you manage your own physical infrastructure. Provisioning infrastructure through code in the cloud — which is a system referred to as Infrastructure as Code (IaC) — enables you to create templates and repeatable processes. When you track changes to your infrastructure code through source control, you permit your team to operate seamlessly and track changes. IaC is much more repeatable and automated — not to mention faster — than having engineers click around a portal.

WARNING

Even instructions on the portal aren't foolproof. You risk making small, yet significant, changes to infrastructure setup if you consistently build the same setup through the portal rather than a YAML file.

Taking your DevOps culture to the cloud

People often speak about DevOps and cloud computing as if they are intertwined and, in many ways, they are. Be aware, however, that you can adopt DevOps — or begin to transform your engineering organization — without going all in on the cloud. It's perfectly reasonable that you first establish the standards, practices, and processes for your team before you shift your infrastructure to a cloud provider.

Although people speak as though everyone is already on the cloud, you are still on the cutting edge of the shift to the cloud. Cloud providers are becoming more robust by the day, and engineering companies are slowly transitioning their self-hosted services to the cloud. With that in mind, an organization seeking to adopt DevOps would be wise to strongly consider utilizing the services of a major cloud provider.

I would never call the cloud a NoOps solution, but I do call it OpsLite. Cloud services often abstract complex operations architecture in a way that makes that architecture more friendly to developers and empowers them to take more ownership of their components. If you've ever grumbled that developers should be included in an on-call rotation, you're right — they should be. Including developers in the on-call rotation is a great way to ramp up their knowledge of deploying code as well as managing and provisioning the infrastructure on which their services run. This reduces operational overhead and frees up the time of operations specialists to work on proactive solutions.

Learning through adoption

If your team is capable of adopting DevOps and shifting toward utilizing cloud computing at the same time, you can use these shifts as learning opportunities for both developers and operations folks.

While your team shifts to the cloud, developers have the opportunity to familiarize operations specialists with code — perhaps even specific languages — and source control, and operations folks can teach developers about infrastructure. When both groups are both the experts and the newbies, neither group has to deal much of an ego-damaging transfer of knowledge.

The trust, rapport, and healthy dynamic that emerge from these interactions will galvanize your team and last much longer than the immediate work took. In many ways, you're reinforcing your DevOps culture through tooling your DevOps practice.

Benefitting from cloud services

Modern operations is changing and evolving. Your competitors are already adopting new ways of innovating faster and accelerating their software delivery life cycles.

Cloud computing represents a big shift from the traditional way businesses think about IT resources. By outsourcing much of your infrastructure and operations requirements to a cloud provider, you reduce overhead and free your team to focus on delivering better software to your users.

Here are six common reasons organizations are turning to cloud computing services:

- **Improving affordability:** Cloud providers allow you to select only the services you need, when you need them. Imagine if you could access cable TV but pay for only the channels you watch. You'd love that, wouldn't you? I would! Cloud providers do just that while also providing you with the most up-to-date computing hardware housed in physically secure data centers.
- **Automating deployments:** Changes to the system — deployments — are the most common contributors of outages or service disruptions. Cloud providers make releasing code an automated, repeatable process, significantly decreasing the probability of making mistakes in manual releases and introducing bugs. Automated deployments also enables developers to release their own code. Ultimately, automated deployments simplify the process while reducing site downtime and reactionary triaging in production.
- **Accelerating delivery:** The cloud reduces friction along nearly every phase of the software delivery life cycle. Although setup is required, it often takes no more than double the time required to do the process manually, and you have to set up a service or process only once. Accelerated delivery gives you a ton of flexibility.

» **Increasing security:** Cloud providers make security part of their offering. Microsoft Azure, Amazon Web Services (AWS), and Google Cloud Platform (GCP) meet different compliance standards and provide policies, services, and controls that will help you reinforce your system's security. In addition, if you utilize a deployment pipeline tool within the cloud, you can add security checks before new code is released to an environment, thereby reducing the possibility of security vulnerabilities.

» **Decreasing failure:** Through cloud build and release pipelines, your team is capable of creating automated tests to confirm functionality, code quality, security, and compliance of any code introduced into your systems. This capability decreases the possibility of bugs while also reducing the risk of problematic deployments.

» **Building more resilient and scalable systems:** The cloud allows organizations to scale up, scale out, and increase capacity within seconds. This elastic scaling enables spinning up compute and storage resources as needed, no matter where in the world your users interact with your product. This approach permits you to better serve your customers and more efficiently manage infrastructure costs.

Cumulus, Cirrus, and Steel: Types of Clouds

No, the heading for this section doesn't refer to the material of the server rack. With the cloud, you don't have to worry about that anymore! In the realm of cloud providers, there are three types of clouds from which to choose: public, private, and hybrid. Hybrid, as you assume, is the combination of public and private clouds. Each of the three options has its benefits and risks, which I discuss in this section.

Public cloud

The most common — and widely used — cloud is public. This type of cloud is provided by a third-party vendor. These vendors provide resources like virtual machines, containers, and storage for engineers to use. In a public cloud, the provider owns and manages all infrastructure. You can manage your access to those services through a portal, CLI commands, or APIs.

Public clouds are by far the most prevalent and relevant to DevOps. You accrue almost no overhead or up-front costs. You pay only for what you use, and you can scale up and down throughout the day at will.

But here's the catch with the public cloud: It has multiple tenants, which is what users are called. Tenants share hardware, storage, and networking with other users. Resources, interests, and concerns are logically separated but are computed on the same hardware.

The main advantages of a public cloud are lowered costs, lack of server maintenance, extremely flexible and capable scalability, and high reliability because of a large network of servers.

TECHNICAL STUFF

Availability is often measured in 9s. If a vendor claims a service has "5 9s" availability, it promises 99.999 percent uptime. And to achieve that, you need $n + 2$ resources. Any application or service with 99.999 percent availability needs to exist on three physical resources. Why? You have to allow for one machine to be down for scheduled maintenance, which leaves you redundancy if one of the remaining machines goes down because of an unforeseen issue. If you had only one machine available during maintenance, you would have no room for service disruptions.

Private cloud

A private cloud offers resources like a public cloud but for use exclusively by a single business. The data can be hosted from the company's data center or through a third-party vendor. Only one user can access all services and infrastructure. No hardware is shared, and the private cloud eliminates (rather arbitrary) security and compliance concerns for companies with extremely specific requirements, including governments and banks. The three major public cloud providers specifically offer government solutions that comply with various standards.

Private clouds are more expensive and can require maintenance, but they do permit more flexibility in customizing the cloud environment. You can experience excellent security while still benefitting from the high scalability of cloud computing.

Hybrid cloud

A hybrid approach is just what it sounds like: an amalgam of the two other options, public and private. A potential hybrid solution can include an on-premise data center, hosted private clouds, and public cloud resources so that companies can benefit from all the positive aspects of every method.

If you're thinking that you also get all the negative aspects of every method as well, you're right. But hybrid works if you partition your services by volume and security need. You can host your email on a public cloud while storing confidential financial data on storage provided by a private cloud.

But the most interesting use of a hybrid cloud is in its approach to a DevOps transformation. If you're currently maintaining your own physical infrastructure, transitioning your services to the cloud will take some time. Adopting a hybrid cloud computing strategy removes the time-sensitive stress from your team, allowing them to make sure that they do the transition right, not just quickly.

Cloud as a Service

Typically, cloud services fall into three categories of service: Infrastructure as a Service (IaaS), Platform as a Service (PaaS), and Software as a Service (SaaS). These services basically correspond to the layer of a traditional tech stack they fall into. You can connect each of these service categories to build a full cloud computing stack and link various cloud services together.

TECHNICAL STUFF

Despite the risk of vendor lock-in — getting stuck with a cloud provider because moving is too expensive or painful — choosing a single cloud provider has its benefits. Azure, AWS, and GCP all design their services to work the most seamlessly with the other services within the provider's portfolio.

Infrastructure as a Service

Infrastructure as a Service (IaaS) is the simplest and most straightforward category of service in the cloud. IaaS provides rented IT infrastructure — low-level network infrastructure via abstracted APIs. You can spin up servers and VMs, storage, backups, and networks. Every service is set up to be pay-as-you-go. You pay for only the resources you use.

Almost all IaaS providers also offer Platform as a Service (PaaS) and Software as a Service (SaaS). Although the user doesn't control the underlying cloud infrastructure, they can manage and control everything that sits on it, including the operating system and particular networking components.

Cloud providers provide these resources elastically from large pools of hardware in secured data centers throughout the world.

Platform as a Service

The platform services that cloud providers offer cover most things you traditionally think of as operations-focused, minus the hardware. Platform cloud services include environments such as development (DEV), quality assurance (QA), user acceptance testing (UAT), staging, and production (PROD). The production environment is exposed to users, but the staging environment provides developers with the opportunity to test their code before it reaches its final release to customers in production.

Platform as a Service (PaaS) is designed to increase the speed at which engineers develop, test, and release their code. With PaaS, developers can develop, test, release, and maintain their applications despite having little to no knowledge about underlying infrastructure. PaaS abstracts servers, storage, databases, middleware, and network resources.

REMEMBER

If you've ever heard someone say that DevOps is leaving Ops behind, PaaS is likely what they're referring to. Many of the tools in this category are targeted toward developers and enable them to act as an operations person. PaaS tools emphasize code because it's automated, controlled, and trackable, not because it seeks to eliminate operations specialists from the delivery life cycle.

The main advantage of PaaS is not having to deal with the complex nature of infrastructure. If you're a developer, you're free to do what you do best — develop and release software. If you're an operations person, PaaS eliminates unnecessary and repeatable toil so that you can focus on solving more interesting and more complex problems.

Going serverless is a concept that you can consider to be part of PaaS. It has reached an adoption rate that merits some specific attention. Going serverless requires servers — surprise! — but represents services that enable functionality without the requirement of server management. With serverless functions, the cloud provider manages much more of the process, including setup and resource management, which allows you to take advantage of scalable and typically event-driven features. Resources are only allocated when a specific function is triggered.

Software as a Service

Software as a Service (SaaS) refers to a hosted and managed application that provides a service. The application is typically accessible through any device. Cloud providers also offer SaaS functionality. The software is licensed and accessed through a subscription model.

Examples of SaaS include TurboTax, Microsoft Office, Slack, Concur, Adobe Creative Suite, Camtasia, Dropbox, and Monosnap. You likely use many more SaaS applications than you even realize. Only when you really start to consider it do you realize just how much functionality is abstracted by SaaS in your everyday life.

Choosing the Best Cloud Service Provider

Selecting a cloud service provider isn't an easy choice. GCP, AWS, and Azure have more in common than they do apart. Often, your decision depends more on your team's comfort level with a particular cloud provider or your current stack more than the cloud provider itself. After you've decided to move to the cloud, the next decision is to decide on a cloud provider. Here are some things to consider when evaluating cloud providers:

- **Solid track record:** The cloud you choose should have a history of responsible financial decisions and enough capital to operate and expand large data centers over decades.
- **Compliance and risk management:** Formal structure and established compliance policies are vital to ensure that your data is safe and secure. Ideally, review audits before you sign contracts.
- **Positive reputation:** Customer trust is absolutely key. Do you trust that you can rely on this cloud provider to continue to grow and support your evolving needs?
- **Service Level Agreements (SLAs):** What level of service do you require? Typically cloud providers offer various levels of uptime reliability based on cost. For example, 99.9 percent uptime will be significantly cheaper than 99.999 percent uptime.
- **Metrics and monitoring:** What types of application insights, monitoring, and telemetry does the vendor supply? Be sure that you can gain an appropriate level of insight into your systems in as close to real-time as possible.

Finally, ensure the cloud provider you choose has excellent technical capabilities that provide services that meet your specific needs. I go into specifics of cloud offerings in the section, "Finding Tools and Services in the Cloud." Generally, look for

- Compute capabilities
- Storage solutions

- Deployment features
- Logging and monitoring
- Friendly user interfaces

You should also confirm the capability to implement a hybrid cloud solution in case you need to at some point, as well as to make HTTP calls to other APIs and services.

The three major cloud providers are Google Cloud Platform (GCP), Microsoft Azure, and Amazon Web Services (AWS). You can also find smaller cloud providers and certainly a number of private cloud providers, but the bulk of what you need to know comes from comparing the public cloud providers.

Amazon Web Services (AWS)

As do the other major public cloud providers, AWS provides on-demand computing through a pay-as-you-go subscription. Users of AWS can subscribe to any number of services and computing resources. Amazon is the current market leader among cloud providers, holding the majority of cloud subscribers. It offers a robust set of features and services in regions throughout the world. Two of the most well-known services are Amazon Elastic Compute Cloud (EC2) and Amazon Simple Storage Service (Amazon S3). As with other cloud providers, services are accessed and infrastructure is provisioned through APIs.

Microsoft Azure

Before Microsoft launched this cloud provider as Microsoft Azure, it was called Windows Azure. Microsoft designed it to do just what the name implies — serve as a cloud provider for traditionally Windows IT organizations. But as the market became more competitive and Microsoft started to better understand the engineering landscape, Azure adapted, grew, and evolved. Although still arguably less robust than AWS, Azure is a well-rounded cloud provider focused on user experience. Through various product launches and acquisitions — notably GitHub — Microsoft has invested heavily in Linux infrastructure, which has enabled it to provide more robust services to a wider audience.

Google Cloud Platform (GCP)

The Google Cloud Platform (GCP) has the least market share of the three major public cloud providers but offers a substantial set of cloud services throughout nearly two dozen geographic regions. Perhaps the most appealing aspect of GCP is that it offers users the same infrastructure Google uses internally. This infrastructure includes extremely powerful computing, storage, analytics, and machine learning services. Depending on your specific product, GCP may have specialized tools that are lacking (or less mature) in AWS and Azure.

Finding Tools and Services in the Cloud

Literally hundreds of tools and services are at your disposal through the major cloud providers. Those tools and services are generally separated into the following categories:

- Compute
- Storage
- Networking
- Resource management
- Cloud Artificial Intelligence (AI)
- Identity
- Security
- Serverless
- IoT

Following is a list of the most commonly used services across all three of the major cloud providers. These services include app deployment, virtual machine (VM) management, container orchestration, serverless functions, storage, and databases. I include additional services such as identity management, block storage, private cloud, secrets storage, and more. It's far from an exhaustive list but can serve as a solid foundation for you as you begin to research your options and get a feel for what differentiates the cloud providers.

- **App deployment:** Platform as a Service (PaaS) solution for deploying applications in a variety of languages, including Java, .NET, Python, Node.js, C#, Ruby, and Go
  - **Azure:** Azure Cloud Services
  - **AWS:** AWS Elastic Beanstalk
  - **GCP:** Google App Engine

- **Virtual machine (VM) management:** Infrastructure as a Service (IaaS) option for running virtual machines (VMs) with Linux or Windows
  - **Azure:** Azure Virtual Machines
  - **AWS:** Amazon EC2
  - **GCP:** Google Compute Engine
- **Managed Kubernetes:** Enables better container management via the popular orchestrator Kubernetes
  - **Azure:** Azure Kubernetes Service (AKS)
  - **AWS:** Amazon Elastic Container Service (ECS) for Kubernetes
  - **GCP:** Google Kubernetes Engine
- **Serverless:** Enables users to create logical workflows of serverless functions
  - **Azure:** Azure Functions
  - **AWS:** AWS Lambda
  - **GCP:** Google Cloud Functions
- **Cloud storage:** Unstructured object storage with caching
  - **Azure:** Azure Blob Storage
  - **AWS:** Amazon S3
  - **GCP:** Google Cloud Storage
- **Databases:** SQL and NoSQL databases, on demand
  - **Azure:** Azure Cosmos DB
  - **AWS:** Amazon Relational Database Service (RDS) and Amazon DynamoDB (NoSQL)
  - **GCP:** Google Cloud SQL and Google Cloud BigTable (NoSQL)

As you explore the three major cloud providers, you notice a long list of services. You may feel overwhelmed by the hundreds of options at your disposal. If, by chance, you can't find what you need, the marketplace will likely provide something similar. The marketplace is where independent developers offer services that plug into the cloud — hosted by Azure, AWS or GCP. Table 21-1 lists additional services provided by most, if not all, cloud providers.

TABLE 21-1 Common Cloud Services

Service Category	Functionality
Block storage	Data storage used in storage-area network (SAN) environments. Block storage is similar to storing data on a hard drive.
Virtual Private Cloud (VPC)	Logically isolated, shared computing resources.
Firewall	Network security that controls traffic.
Content Delivery Network (CDN)	Content delivery based on the location of the user. Typically utilizes caching, load balancing, and analytics.
Domain Name System (DNS)	Translator of domain names to IP addresses for browsers.
Single Sign-On (SSO)	Access control to multiple systems or applications using the same credentials. If you've logged into an independent application with your Google, Twitter, or GitHub credentials, you've used SSO.
Identity and Access Management (IAM)	Role-based user access management. Pre-determined roles have access to a set group of features; users are assigned roles.
Telemetry, monitoring, and logging	Tools to provide application insights on performance, server load, memory consumption, and more.
Deployments	Configuration, infrastructure, and release pipeline management tools.
Cloud shell	Shell access from a command-line interface (CLI) within the browser.
Secrets storage	Secure storage of keys, tokens, passwords, certificates, and other secrets.
Message Queues	Dynamically scaled message brokers.
Machine Learning (ML)	Deep learning frameworks and tools for data scientists.
IoT	Device connection and management.

6 The Part of Tens

IN THIS PART . . .

Gain a clear grasp of the top ten reasons that you and your organization benefit from adopting DevOps.

Be prepared for the biggest challenges that can arise as you undertake your DevOps transformation.

» **Competing with the best through DevOps methodologies**

» **Continuously improving your systems and processes**

Chapter 22
Top Ten (Plus) Reasons That DevOps Matters

This chapter presents the key points to know about how DevOps benefits your organization. Use it as a reference to help you persuade your colleagues or to reinforce your understanding of why you chose to go the DevOps route when the road gets bumpy.

Accepting Constant Change

The tech landscape is an ever-changing environment. Some languages evolve and new ones are created. Frameworks come and go. Infrastructure tooling changes to meet the ever-growing demands for hosting applications more efficiently and delivering services more quickly. Tools continue to abstract low-level computing to reduce engineering overhead.

The only constant is change. Your ability to adapt to that change will determine your success as an individual contributor, manager, or executive. Regardless of the role you currently fill at your company or hope to eventually play, it is vital to adapt quickly and remove as much friction from growth as possible. DevOps enables you to adapt and grow by improving communication and collaboration.

Embracing the Cloud

The cloud isn't the future; it's now. Although you may still be transitioning or not yet ready to move, realize that the cloud is the way forward for all but a few companies. It gives you more flexibility than traditional infrastructure, lowers the stress of operations, and (usually) costs significantly less because of a pay-as-you-go pricing structure. Public, private, and hybrid clouds give you endless possibilities to run your business better. The ability to spin up (launch) resources within minutes is something most companies have never experienced prior to the cloud.

This agility provided by the cloud goes hand in hand with DevOps. Omri Gazitt from Puppet, a company focused on automation and configuration management, put it best: "As organizations move to the cloud, they are revisiting their core assumptions about how they deliver software." With the cloud, APIs connect every service, platform, and infrastructure tool so that you can manage your resources and application seamlessly. As you migrate to the cloud, you can reevaluate past architecture decisions and slowly transition your application and system to be cloud-native, or designed with the cloud in mind.

Hiring the Best

Because of increased demand, great engineers are scarce. There simply aren't enough engineers to fill all the jobs currently open or to meet market demand over the next decade and beyond. Although finding engineers can be difficult, it's not impossible, especially if you focus on discovering engineers who embrace curiosity and aren't afraid to fail. If you implement DevOps in your overall engineering culture, you can level up engineers and train them in the methodology and technology that supports continuous improvement.

It's difficult to measure potential in an interview. I believe talent whispers. The most talented engineers I've ever met aren't gregarious or braggarts; they let their work speak for them. DevOps enables you to listen more closely to the personal and professional interests of the engineers you interview. I choose candidates based on their level of curiosity, communication skills, and enthusiasm. Those qualities can see your team through the troughs of fear, uncertainty, and doubt. They can carry the team through hard decisions, made within constraints, in their attempt to solve difficult problems.

You can teach someone a skill, but teaching someone how to learn is an entirely different matter. The learning culture you create in your DevOps organization enables you to prioritize a growth mindset over technical prowess. In DevOps, hiring for the team is critical. Every individual is a piece of a whole, and the team must have balance holistically. Achieving this balance means that sometimes you don't hire the "best" engineer, you hire the best engineer for *the team.*

When you hire for the team you can, like draft horses yoked together, pull more weight than you could individually. With DevOps, you can multiply the individual components of your team and, as a whole, create a powerhouse of a team.

Staying Competitive

The yearly "State of DevOps Report" released by DevOps Research and Assessment (DORA) makes it clear: Companies across the world are using DevOps to adjust their engineering practices and are reaping the benefits. They see increases in engineering production and reductions in cost. With DevOps, these companies are shifting from clunky processes and systems to a streamlined way of developing software focused on the end user.

DevOps enables companies to create reliable infrastructure and utilize that infrastructure to release software more quickly and more reliably. The bottom line is this: High-performing organizations use DevOps, and they're crushing their competition by increasing their deployment frequency and significantly decreasing their failures that occur because of changes in the system. If you want to compete, you must adopt the methodologies described in this book. Maybe not all of them, and definitely not all at one time — but the time to wait and see whether DevOps is worthwhile has passed.

Solving Human Problems

Humans have reached a point in our evolution at which technology is evolving faster than our brains. Thus the greatest challenges humans face are due to human limitations — not the limitations of our software or infrastructure. Unlike other software development methodologies, DevOps focuses holistically on your socio-technical system.

Embracing DevOps requires a shift in culture and mindset. But if you achieve a DevOps culture and mindset, you and your organization reap almost limitless benefits. When engineers are empowered to explore, free of the pressure and fear of failure, amazing things happen. Engineers discover new ways to solve problems. They approach projects and problems with a healthy mindset and work together more fluidly, without needless and negative competition.

Challenging Employees

DevOps accelerates the growth of individual engineers as well as that of the engineering team as a whole. Engineers are smart people. They're also naturally curious. A great engineer who embraces a growth mindset needs new challenges after mastering a particular technology, tool, or methodology or they often feel stagnant. They need to feel as if their brain and skill sets are being stretched — not to the point of being overwhelmed or stressed, but enough to feel that they're growing. That is the tension described by Dan Pink in *Drive*. If you can strike that balance, your engineers will thrive — as individuals and as a team.

The methodology of DevOps promotes T-shaped skills, which means that engineers specialize in one area with deep knowledge and have a broad understanding of many other areas. This approach allows engineers to explore other areas of interest. Perhaps a Python engineer has an interest in cloud infrastructure, for example. No other engineering methodology permits and encourages engineers to explore as much as DevOps does, and it's a huge contributor to hiring and retaining talent.

Bridging Gaps

One of challenges of modern technology companies is this gap between the needs of the business and the needs of engineering. In a traditional company, with traditional management strategies, a natural friction exists between engineering and departments like marketing, sales, and business development. This friction stems from a lack of alignment. Each department is measured by different indicators of success.

DevOps seeks to unify each department of a business and create a shared understanding and respect. That respect for each other's jobs and contributions is what allows every person in the company to thrive. It removes the friction and improves acceleration.

Think about a team of sled dogs. If each dog is moving in separate directions, the sled goes nowhere. Now imagine the dogs working together, focused on moving forward — together. When you lack friction internally, the only challenges you face are external, and external challenges are almost always more manageable than internal strife.

Failing Well

Failure is inevitable. It's simply unavoidable. Predicting every way in which your system can fail is impossible because of all the unknowns. (And it can fail spectacularly, can't it?) Instead of avoiding failure at all costs and feeling crushed when failure does occur, you can prepare for it. DevOps prepares organizations to respond to failure, but not in a panicky, stress-induced way.

Incidents will always involve some level of stress. At some point along your command structure, an executive is likely to scream at the money being lost during a service outage. But you can reduce the stress your team experiences by using failure as a way of learning and adapting your system to become more resilient. Each incident is an opportunity to improve and grow, as individuals and as a team.

DevOps embraces kaizen, the art of continuous improvement. When your team experiences flow in their work, they can make tiny choices every day that contribute to long-term growth and, ultimately, a better product.

Continuously Improving

I talk a lot about acceleration and continuous improvement throughout this book. Use the visualization of a never-ending cycle from Chapter 6 and apply it to your organization. The cycle shouldn't invoke fears through thoughts of Sisyphus, pushing a boulder up a hill for all eternity. Instead, think of this cycle as movement, like a snowball rolling downhill, gathering momentum and mass.

As you adopt DevOps and integrate more and more of its core tenets into your everyday workflow, you'll witness this acceleration first-hand. The cycle of continuous improvement should always center around the customer. You must continuously think about the end user and integrate feedback into your software delivery life cycle.

Fundamental to this cycle is CI/CD (explained in Chapter 11). Adopting CI/CD isn't an all-or-nothing requirement of DevOps; instead, it's a slow process of implementation. You should focus on mastering continuous integration first. Encourage engineers to share code freely and merge code frequently. This approach prevents isolation and silos from becoming blockers in your engineering organization.

After your organization masters continuous integration, move on to continuous delivery, the practice of automating software delivery. This step requires automation because code will move through multiple checks to ensure quality. After all your code is secure and accessible in a source code repository, you can begin to implement small changes continuously. Your goal is to remove manual barriers and improve your team's ability to discover and fix bugs without customer impact.

Automating Toil

Acceleration and increased efficacy are at the core of the DevOps methodology. By automating labor-intensive manual processes, DevOps frees engineers to work on projects that make the software and systems more reliable and easily maintained — without the chaos of unexpected service interruptions.

Site reliability engineering (SRE) deals with toil, which is the work required to keep services up and running but is manual and repetitive. Toil can be automated and lacks long-term value. Perhaps most important of all, toil scales linearly, which limits growth. Note that toil doesn't refer to the overhead of administrative necessities such as meetings and planning. This type of work, if implemented with a DevOps mentality, is beneficial to the long-term acceleration of your team.

One of the core tenets of tooling your DevOps practice is automation. You can automate your deployment pipeline to include a verbose test suite as well as other gates through which code must pass to be released. In many ways, SRE is the next logical step in the evolution of DevOps and should be your next step after you and your organization master the core concepts of DevOps and implement the practice in your team.

Accelerating Delivery

The software delivery life cycle has evolved from the slow and linear Waterfall process to an agile and continuous loop of DevOps. You no longer think up a product, develop it fully, and then release it to customers, hoping for its success. Instead, you create a feedback loop around the customer and continuously deliver iterative changes to your products. This connected circuit enables you to continuously improve your features and ensure that the customer is satisfied with what you're delivering.

When you connect all the dots of this book and fully adopt DevOps in your organization, you watch as your team can deliver better software faster. The changes will be small at first, just like the changes you release. But over time, those seemingly insignificant changes add up and create a team that accelerates its delivery of quality software.

» **Forgetting to measure**

» **Fearing failure**

» **Implementing DevOps too rigidly**

Chapter 23
Top Ten DevOps Pitfalls

Fostering a DevOps culture and selecting tools to support your DevOps approach will benefit your organization. It galvanizes your engineering team and focuses your product development on your customer.

However, any time you attempt to make a massive change to the undercurrent of your organization, you face challenges and have to deal with setbacks. As you transform to DevOps, you'll discover unique speed bumps for you and your team to get over. Although I can't possibly predict every obstacle you'll face, this chapter can prepare you for the ten most common DevOps pitfalls. Remember that however you approach your DevOps practice, your priorities should remain focused on people, process, and technology — in that order.

Deprioritizing Culture

More than anything else, DevOps is a cultural movement. The culture you build at your organization will make or break your DevOps practice. Your DevOps culture must emphasize collaboration, trust, and engineering empowerment. If you nail automation but miss those cultural components, you will likely fail.

In truth, tooling doesn't matter that much. The tools you have at your disposal are more similar than not. Although the problems they solve are important, none of those problems can compare to the nearly endless frustration of trying to unite developers and operations folks — as well as other teams, like security — in a traditional engineering organization.

DevOps seeks to galvanize engineers (as well as business groups). It creates a foundation on which everyone can learn, share, and grow. That personal acceleration will fuel your entire engineering organization to create better software, faster. The engineers you have on your team are the most valuable asset you have. Treat them well by giving them respect and the room to do what they do best — engineer solutions.

Leaving Others Behind

Making the case internally for DevOps will determine the type of foundation you build for your culture. Look for fertile soil. If you move too quickly and don't convince key people of the importance of a DevOps transformation, people will watch your movements with skepticism and leap at the first opportunity to show everyone you're wrong. That is not a fun position to be in, and you never want to start this journey with people waiting for you to fail.

To be successful, you need everyone on board, even the naysayers and skeptics. Engineers can be skeptical. After a decade or two in this industry, they've seen a lot of ideas and new approaches come and go. They can easily shrug off DevOps as "just another failed approach" to the same old problems. And if you implement it poorly, DevOps will indeed be just another failed approach. You and your team must persuade others of the potential and take action in ways that invite everyone to the table.

I recommend convincing executives with data and the potential for accelerated software delivery. But engineers need to know how DevOps will make their jobs more enjoyable. Show them how DevOps aligns with business needs and reduces friction along the software delivery pipeline. Just be sure not to oversell the concept. DevOps is not a silver bullet and requires intense work at the beginning to ensure that the team creates a learning culture in which engineers are free to make mistakes and grow.

After you reach an event horizon where enough people believe in DevOps, you can proceed with the knowledge that you have the support of your organization and the people within it.

Forgetting to Align Incentives

If you don't set out to align incentives with what you expect from certain teams or specific engineers, more challenges arise. The real tool of DevOps, if you can master it, is empowerment. You want to empower your engineers to do their job well, free from interference. You hired talented engineers, so trust their ability to fulfill their responsibilities.

For example, when developers serve on an on-call rotation, some organizations frame it as a bit of a punishment. "You built it, you support it" doesn't exactly fill people with happy feelings. Instead, it feels like just another form of siloed responsibility. But a humane and evenly distributed on-call rotation not only empowers developers to take ownership of their work, but it also creates learning opportunities for the entire team.

In DevOps, you don't punish engineers for imperfect work; instead, you share responsibility and cultivate an organization that values learning and empowers everyone to be curious as well as participate in areas of tech in which they're less familiar.

Aligning incentives and creating opportunities for collaboration drives your goal of improving your products and better serving your customers. If everyone is aligned toward the goal of creating amazing services for your customers, you will see the group begin to galvanize.

Keeping Quiet

DevOps is the antithesis of secrets and backroom negotiations. Instead, it lays everything out on the table and forces you to trust the integrity of the people in your organization. When you first introduce open communication, conflict may seem to increase. It doesn't. Instead, you're simply seeing the friction points for the first time. Instead of leaving conflict to brew beneath the surface, people feel safe enough to raise their concerns and express their opinions.

An important aspect of open communication is to keep it going throughout the entire product life cycle — from ideation to production. You must include engineers in planning discussions, architecture decisions, development progress updates, and deployments. Although this emphasis on communication creates more verbose discussions, it also enables engineers to have visibility outside of their core area of expertise, which in turn empowers them to advise others while equipped with the context necessary to make sound decisions.

Keep the customer — and what they expect from the product you're building — at the center of every discussion and decision. If you stay aligned on that goal, you're sure to move forward together as one unit.

Forgetting to Measure

Measuring your progress is crucial to DevOps success. It lends you validation when making the argument for DevOps to doubting stakeholders, helps you convince holdout executives, and reminds your engineering team how much they've accomplished.

Before you make a single change, create a baseline. Choose a small set of data you want to track through your entire process. This data informs your decisions and serves as fuel to continue pushing when you hit setbacks. Potential measurements include:

- **Employee satisfaction:** Do your engineers love working at your organization?
- **Monthly recurring revenue (MRR):** How much money are you making from customers?
- **Customer tickets:** How many bugs are reported by your customers?
- **Deployment frequency:** How many deployments do you have every week or month?
- **Mean time to recovery (MTTR):** How long does take to recover from a service disruption?
- **Service availability:** What is the uptime of your application? Are you hitting your current service-level agreements?
- **Failed deployments:** How many releases cause service disruptions? How many have to be rolled back?

Micromanaging

One of the quickest ways to undermine your engineers is to micromanage their work. Dan Pink, author of the book *Drive,* believes that motivation at work is driven by three factors:

- Autonomy
- Mastery
- Purpose

Extrinsic motivators like high salaries, bonuses, and stock options may work in the short-term, but long-term job satisfaction depends more on personal and professional growth. You want your engineers to exist in the tension of feeling highly challenged but not overwhelmed by stress. That sweet spot is different for every person. If you can evoke someone's passion, they're sure to work enthusiastically.

Trust is critical to DevOps organizations. You must trust your colleagues, peers, engineers, managers, and executives. You must also trust the roles and responsibilities of the various departments in your organization — which isn't to say that you will never have conflict. Of course moments of friction will happen between human beings. But minimizing those moments and enabling healthy conflict resolution is what distinguishes DevOps-focused engineering teams from their competition.

Changing Too Much, Too Fast

Many teams make too many changes too quickly. Humans don't like change. (I certainly don't.) Although DevOps is beneficial over the long term, quick changes to the normal way of doing things can be jarring to engineers.

One failing of DevOps is that it implies that everyone lives in the greenfield (new software) with rainbows and unicorns. It can sound like, "If only you can get your team to work together, software development will be easy!" That's not true. Software engineering is hard and will always be hard. That's one thing most engineers *like* about it. You enjoy a challenge. But challenges should be stimulating, not stressful.

DevOps doesn't aim to remove all the intellectual challenges of engineering. Instead, it offers to minimize the friction between humans so that everyone can focus on their work. If you attempt to make too many changes too quickly, you can find yourself in the middle of an all-out revolt — Mutiny on the Binary.

Choosing Tools Poorly

Although I deprioritize tooling in DevOps — and rightfully so — tooling *is* still a factor. Even the least important aspect of DevOps contributes to your overall success. The tools you select should solve the problems your engineering team experiences, but should also align with the style, knowledge, and comfort areas of your existing team.

Don't be afraid to try several solutions and see which one fits the best. Dedicating a few weeks to a minimum viable product (MVP) or proof of concept (POC) to test a tool is well worth the effort. Even if you end up throwing it away, "wasting" the engineering resources is preferable to going all-in on a particular technology only to find out a year later that it's not a good fit.

Fearing Failure

Failing fast is a short way of saying you should constantly be iterating to identify problems early in the process without spending a ton of time and money. I discuss failing fast more in Chapter 16. It's something that a lot of people in tech talk about and few actually implement because it requires rapid iteration in an environment in which mistakes have a small blast radius and are easily corrected. Too often, companies claim a fail-fast mentality and instead fire the first engineer to delete a production database. (As if any engineer out there has never deleted a production database . . . .)

In the context of DevOps, however, you're better off failing well than failing fast. Failing well implies that you have monitoring in place to alert you to potential problems long before the situation impacts customers. Failing well also implies that you've designed your system in a segmented way that prevents one service that's falling over from cascading into a systemic outage. But organizations that fail well go one step further as well: They don't blame people. Instead, they look for failures in systems and processes.

Kaizen is the Japanese word for continuous improvement. In DevOps, kaizen means to continuously improve your processes. It's not some sexy transformation that has a beginning and an end. The goal isn't to go from zero to perfect. Instead, DevOps encourages working slowly and gradually toward making one thing better, every day. If you leave work each evening knowing that just one small aspect of work is better because of you, wouldn't you feel satisfied? I would and do, and I'm willing to bet that a lot of engineers feel the same.

Instead of attempting to avoid failure at all costs, DevOps insists on a growth mindset. Failure isn't a marker of stupidity or poor preparation. It's a marker of growth and a necessary step in innovation. Innovation is an outcome that I hope you're willing to pursue, even if it means that you occasionally fail.

Being Too Rigid

DevOps is not prescriptive, and that's both the best and worst thing about it. DevOps would be so much easier to implement if I could give you ten actions to take to achieve DevOps nirvana. I wish I could! But humans don't work that way, and groups of humans — such as on engineering teams and in large organizations — create even more complexities that need to be addressed.

Although no blueprint for building a DevOps organization exists, you are empowered to tailor the methodology to practices that work for you and your team. You know your organization, and I encourage you to think out of the box when applying the fundamentals. Some of the things in DevOps will fit you perfectly. Others will feel like wearing a jacket that's just one size too small. That's okay.

You're going to make mistakes. No one is perfect. But if you let go a bit, empower your engineers, and trust your team, you will see awesome outcomes. Just get started. And remember: Invite everyone to the table, measure your progress, prioritize culture over technology, and empower your engineers to do what they do best.

Index

Numbers

A

B

C

D

E

F

G

H

I

J

K

L

M

N

O

P

Q

R

S

T

U

About the Author

Emily Freeman is a technologist and a storyteller who helps engineering teams improve their velocity. She believes the biggest challenges facing developers aren't technical, but human. Her mission in life is to transform technology organizations by creating a company cultures in which diverse, collaborative teams can thrive.

Emily is a Senior Cloud Advocate at Microsoft, and her experience spans both cutting-edge startups and some of the largest technology providers in the world. Her work has been featured in outlets such as Bloomberg and she is widely recognized as a thoughtful, entertaining, and professional keynote speaker. Emily is best known for her creative approach to identifying and solving the human challenges of software engineering. It is rare in the technology industry to find individuals equally adept with code and words, but her career has been defined by precisely that combination.

Emily lives with her daughter in Denver, Colorado.

Dedication

For Clara, my North Star; and for all those who came before me, whose seemingly inconsequential decisions led to this moment of joy and accomplishment.

Acknowledgments

This book was far from a solo effort. I can't possibly thank everyone who gave me small notes of encouragement or who believed in me when I didn't believe in myself. From the bottom of my heart, thank you.

First, to my reader. Thank you for investing your precious time to reading my thoughts on the fundamentals and implementation of DevOps. It's my sincere hope that you're able to walk away from this book feeling empowered to make changes for yourself and your colleagues.

Many thanks to Steven Hayes, who took a chance on a first-time author. This opportunity meant the world to me. To Susan Christophersen, who edited this book diligently. This book is better because of her work. Thank you also to Nicole Forsgren for her amazing foreword (and sage advice) and to Jason Hand for his technical edits.

A special thanks to my parents, Barry and Pamela Freeman, who walked me off the ledge of quitting more times than I can say. Thank you for your love, for teaching me the importance of creative work, and for giving me the gift of empathy. I love you.

I owe much of my success to the authors who have encouraged me along the way. They spent hours acknowledging the pain of writing and cheering me on. Thank you to John Allspaw, David Blank-Edelman, Sarah Drasner, Chad Fowler, Stephen O'Grady, Mike Julian, Gene Kim, Niall Murphy, Erik St. Martin, Mary Thengvall, and James Turnbull.

I want to thank those of you in the community who believed in me long before the thousands of Twitter followers. I was a backend engineer fresh out of code school and your confidence in my potential carried me further than you'll ever know. I owe so much to people like Aaron Aldrich, Kent Dodds, James Governor, Nathen Harvey, Christian Herro, Bridget Kromhout, Ken Mugrage, Corey Quinn, J. Paul Reed, Matt Rogers, Michael Stahnke, Matty Stratton, and Joshua Zimmerman.

I'm convinced I have some of the most stellar personal friends on the planet. I'd be typing all day if I listed everyone, but I want to highlight a few particular people who played an important role in this book. Thank you to Jessica West for being my most trusted friend and confidant. (I don't know what I'd do without you.) Scott Church, thank you for letting me be me without judgement. Melanie Parish, you are my sound advisor. Rachel Stephens, your contagious joy is a gift. Heidi Waterhouse, thank you for checking in on me and grounding me. Chris Short, thank you for your dark humor and candor. Lovisa Svallingson, thank you for giving me unconditional love when I feel empty. Kristin Jones, Amber Rivera, and Mary MacCarthy, you inspire me and keep me afloat.

Publisher's Acknowledgments

Executive Editor: Steve Hayes

Project Manager and Copy Editor: Susan Christophersen

Technical Editor: Jason Hand

Editorial Assistant: Matthew Lowe

Proofreader: Debbye Butler

Production Editor: Siddique Shaik

Cover Image: © Graiki/Getty Images

PERSONAL ENRICHMENT

9781119187790
USA $26.00
CAN $31.99
UK £19.99

9781119179030
USA $21.99
CAN $25.99
UK £16.99

9781119293354
USA $24.99
CAN $29.99
UK £17.99

9781119293347
USA $22.99
CAN $27.99
UK £16.99

9781119310068
USA $22.99
CAN $27.99
UK £16.99

9781119235606
USA $24.99
CAN $29.99
UK £17.99

9781119251163
USA $24.99
CAN $29.99
UK £17.99

9781119235491
USA $26.99
CAN $31.99
UK £19.99

9781119279952
USA $24.99
CAN $29.99
UK £17.99

9781119283133
USA $24.99
CAN $29.99
UK £17.99

9781119287117
USA $24.99
CAN $29.99
UK £16.99

9781119130246
USA $22.99
CAN $27.99
UK £16.99

PROFESSIONAL DEVELOPMENT

9781119311041
USA $24.99
CAN $29.99
UK £17.99

9781119255796
USA $39.99
CAN $47.99
UK £27.99

9781119293439
USA $26.99
CAN $31.99
UK £19.99

9781119281467
USA $26.99
CAN $31.99
UK £19.99

9781119280651
USA $29.99
CAN $35.99
UK £21.99

9781119251132
USA $24.99
CAN $29.99
UK £17.99

9781119310563
USA $34.00
CAN $41.99
UK £24.99

9781119181705
USA $29.99
CAN $35.99
UK £21.99

9781119263593
USA $26.99
CAN $31.99
UK £19.99

9781119257769
USA $29.99
CAN $35.99
UK £21.99

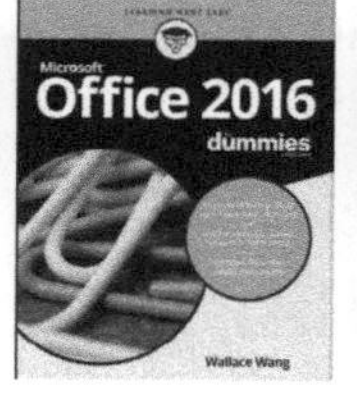

9781119293477
USA $26.99
CAN $31.99
UK £19.99

9781119265313
USA $24.99
CAN $29.99
UK £17.99

9781119239314
USA $29.99
CAN $35.99
UK £21.99

9781119293323
USA $29.99
CAN $35.99
UK £21.99

dummies.com

dummies
A Wiley Brand

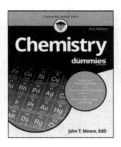

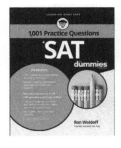

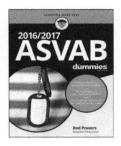

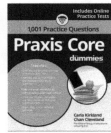